Palgrave Studies in Sub-National Governance

This series explores the formal organisation of sub-national government and democracy on the one hand, and the necessities and practices of regions and cities on the other hand. In monographs, edited volumes and Palgrave Pivots, the series will consider the future of territorial governance and of territory-based democracy; the impact of hybrid forms of territorial government and functional governance on the traditional institutions of government and representative democracy and on public values; what improvements are possible and effective in local and regional democracy; and, what framework conditions can be developed to encourage minority groups to participate in urban decision-making. Books in the series will also examine ways of governance, from 'network governance' to 'triple helix governance', from 'quadruple' governance to the potential of 'multiple helix' governance. The series will also focus on societal issues, for instance global warming and sustainability, energy transition, economic growth, labour market, urban and regional development, immigration and integration, and transport, as well as on adaptation and learning in sub-national government. The series favours comparative studies, and especially volumes that compare international trends, themes, and developments, preferably with an interdisciplinary angle. Country-by-country comparisons may also be included in this series, provided that they contain solid comparative analyses.

Amy Burnett

The Politics of Transition

Innovative Place-Making and Alternative Development Models Under English Localism

palgrave
macmillan

Amy Burnett
Centre for Enterprise, Environment
and Development Research (CEEDR)
Middlesex University
Hendon, London, UK

ISSN 2523-8248 ISSN 2523-8256 (electronic)
Palgrave Studies in Sub-National Governance
ISBN 978-3-031-40768-0 ISBN 978-3-031-40769-7 (eBook)
https://doi.org/10.1007/978-3-031-40769-7

This Palgrave Macmillan imprint is published by the registered company Springer Nature Switzerland AG.
The registered company address is: Gewerbestrasse 11, 6330 Cham, Switzerland

Paper in this product is recyclable.

Ethical Considerations

Participants have been informed about the purpose, methods and intended possible uses of the research. The researcher has no conflicts of interest that affect the ethical soundness of the research.

Abstract

The widespread, global recognition of climate and ecological emergencies indicates that the way we plan for our communities needs to change. For too long, development has compromised low-carbon and nature-sensitive, quality design. Meanwhile, people are turned off by the technocratic rules of the planning process, meaning that the wider community may not always have a seat at the table. Too often we hear the power of developers can undercut local interest. The book aims to contribute to discussions on sustainability transitions (that is, purposeful, meaningful shifts towards sustainable futures), using the vehicles of community-led planning (such as Neighbourhood Planning) or other new forms of citizen–government collaboration in the context of climate and ecological emergencies (e.g., such as the UK Government's 25-Year Environment Plan as well as proposed Citizen Assemblies). This is set within wider debates on 'levelling up' social and economic disparities within English 'regions', just transitions and the contribution local government and communities can take towards 'nature-positive' development.

In this book, we explore how communities are using the planning process to shape where they live in the context of the English localism agenda. This policy agenda, instigated by the Conservative party and solidified in the 2010–2015 Conservative–Liberal Democrat Coalition through the 2011 Localism Act, has been the pinnacle in shaping the rights communities have to influence where they live and local government (in theory at least) to respond to these needs. However, the relevance of this study is more than just the English context.

Localism has also been seen by some as a means to solicit more radical forms of placemaking and place-shaping, able to cast new approaches to a more community-centred, sustainable, resilient and experimental local participatory landscape. Since the localism agenda came into force, the UK has contended with Brexit, heightened climate action and awareness, the COVID-19 pandemic and a rising cost-of-living crisis. These multiple crises are not unique to the UK but the configuration of power at the local level to think creatively about how to cope with them and 'build back better' invites a critical assessment of what legacy the Localism Act has had on the bottom-up capacity to dream, and enact, a more sustainable, inclusive form of prosperity in the context of a 'New Economy'; that is, an economy that embraces environmental issues at its core, fosters inclusive governance and generates innovative and appropriate place-based solutions.

We focus on the interaction between the domains of planning, politics and placemaking to explore how development alternatives can become embedded into the fabric of local (state) governance. We do this by examining the relationship between the social movement the Transition town movement ('Transition') and a town council in England through the lens of the Neighbourhood Development Plan (NDP) process, critically assessing how radical approaches that seek to challenge mainstream planning—and politics—are crafted and contested in, and beyond, the planning process. Notably, has localism enabled grassroots, 'niche' influence on mainstream, 'regime' institutions?

As well as reviewing the potential of localism to facilitate inclusive climate action and the role of planning to effect sustainability transitions, we take an in-depth account of how the English market town of Frome in Somerset came to buck a recent trend of a decline of engagement in politics. Frome's Town Council (FTC) has gone from being a relatively 'normal' town with elected representatives from political parties on the town council to become a strong political and social movement under an independent group, Independents for Frome (IfF). Since 2011, IfF has committed to a progressive agenda that seeks to circumvent, subvert and renegotiate the political terrain. IfF is actively constructing a new political model, drawing on the discourse of localism (a policy directive of the central government) as a means to orchestrate the scaling up of parallel infrastructure in the everyday politics of the town. Notably, under IfF, FTC has established many bold and ambitious initiatives both inside and outside the remit of urban planning—including several pioneering approaches that address local action on climate change and the environment; many of which have roots within Sustainable Frome (the local Transition Initiative, part of the wider *Transition* movement).

The relevance of how these initiatives came about is even more urgent as 83% of county and unitary authorities have declared a climate emergency, although less than 15% have declared an ecological emergency (Inkap, 2021). As of September 2022, 340 councils from the district level and above had made climate emergency declarations (Climate Emergency UK, 2022). There has been a scramble of interest in local-level responses to climate change, but the longevity of the response can broadly mirror political interest and the capacity of local stakeholders to craft and own their response to it. At the very local level, few town and parish councils have declared climate and ecological emergencies—partly because many have struggled with obtaining the confidence to do so, and act on such declarations (NALC, 2021)—though interest to address this by local councils themselves is steadily growing.

In this book, we explore what factors explain local environmental and socio-political innovations and the extent that Frome's experience can be deemed as a sustainability transition towards relocalisation; that is, encouraging localised, reinforcing sustainable feedback loops that support effective community organising, political citizenship and economically resilient places. The *Transition* movement has been key to championing relocalisation to develop 'parallel infrastructure' from business-as-usual, neoliberal societies. We ask whether rights bestowed to local authorities under the Localism Act offer a mechanism to lock in (subversive) sustainability and community-led agendas within mainstream development processes. Notably how 'formalised' policies, practices and spheres of governance relate to 'bottom-up' activities in pursuit of new models of socio-political development in and around the planning process.

Using a multi-level and longitudinal perspective, the case study of Frome can help reveal the intersection between place, power and politics and the relationship planning has to broader sustainable development pathways. Linked to this, the book also explores the role of social movements (such as Transition Initiatives) in clustering, transmuting and channelling networks of alternative socio-political pathways and the extent to which different 'tribes' of development alternatives (such as cohousing and regenerative economic thinking) might be 'coming together' under the guise of New Economy discourse and practices (see Real Economy Lab, 2016).

Key debates in sustainability transitions research are invoked and challenged, such as unpacking grassroots 'niche' and mainstream 'regime'-led drivers and inhibitors for sustainable action, the governance of transition arenas (spaces where sustainable outcomes are actively cultivated) and the notion of transition itself to explore the politics of placemaking dynamics that generate or contest 'sustainable' development alternatives. In particular, the study explores the conditions for the influence of relocalised (New Economy) actors and entities on Frome's placemaking and politics; whether 'niche' protagonists possess and retain this identity when occupying formal spaces of power and the role of the 'regime' in cultivating and contesting or development alternatives and broader shifts in environmental, social and political arrangements.

By exploring placemaking (the act of 'making places'), the research gives a spatially-situated account of the role of place in defining and shaping the selection environment for protagonist actors who seek development alternatives and how these definitions are mediated through and within the institutions of the state. For instance, by revealing the specific entities that provide a foundation for protagonist actors to craft and use for their own (politicised) ends. The findings point to how autonomy and independence justified an orchestration of transition governance, with both emergent and strategic foundations (see Burnett and Nunes, 2021).

The book appeals to academics across several disciplines with theoretical and practical implications for planning theory and practice, political science and democracy in a post-Brexit and post-pandemic recovery context, and the role of social movements in breaking down, or co-creating, institutions. The book is also timely for practitioners seeking a critique of the politics involved in advancing alternative political and development approaches at a local level or hoping for a more in-depth account of Frome's apparent recent successes.

Keywords • Climate and Ecological Emergencies • Frome• Independent Politics • Localism • Neighbourhood Planning • New Economy • Relocalisation • Sustainability Transitions • Planning Transitions Theory • Placemaking Transitions • *Transition* Movement.

References

Burnett, A., & Nunes, R. (2021). Flatpack Democracy: power and politics at the boundaries of transition. Environmental Policy and Governance, 1–14. https://doi.org/10.1002/22t.1931

Climate Emergency UK. (2022). Climate Emergency Declarations. https://data.climateemergency.uk/councils/

Inkap. (2021). Less than 15% of councils have declared an Ecological Emergency. Why the delay? https://www.inkcapjournal.co.uk/less-15-of-councils-have-declared-an-ecological-emergency-why-the-delay/

NALC. (2021). *What can local councils do on climate change*. National Association of Local Councils. https://www.nalc.gov.uk/library/our-work/climate-change/3598-what-can-local-councils-do-on-climate-change-2021/file

Real Economy Lab. (2016). Real Economy Lab website. http://reconomy.org/

Acknowledgement and Funding

Neighbourhood plans are often produced with and supported by the dedication and effort of committed community members as a means to positively shape where they live. This book, I hope, sheds light on the politics of trying to use neighbourhood plans as a tool to encourage sustainable societies. It is dedicated to people everywhere who want to use community planning tools to make where they live places people enjoy now and in the future and offers some insights into understanding the importance, and complexity, of transformative environmental, political and social innovations. Taking a critical view in unpacking how this happens does not undermine my own belief that neighbourhood plans can be powerful tools to do this, together with other, complementary, placemaking strategies. Through the course of writing this book I have also learned a lot about the channels through which a 'greener' politics flows and how social and innovative political movements can rise, and sometimes fall or transmute their energies into unexpected courses. This research has been an exciting pursuit of a fascinating research topic, navigating the complexity of the strands through which Frome's story can be told and its implications for scaling out some of its 'niche' experimentation in environmental politics and planning. I am truly grateful to everyone who has supported me on this path, particularly my family, who have given so much as I have drawn on them for time and space. I am sincerely grateful for all the precious time my PhD supervisors, Dr Richard Nunes and Prof Gavin Parker from the University of Reading's (UoR's) Henley Business School (HBS), and the Examiners, Prof Jane Wills (University of Exeter) and Dr Emma Street (UoR/HBS), who contributed to helping me bring forward an improved thesis and to refine fuzzy ideas. Thank you to the respondents of Frome who shared their fascinating stories, feedback and their patience. I'd also like to thank the people who have met with me to discuss ideas about the measurement and incentivisation of sustainable action and I look forward to these ideas manifesting in some form or another. This includes including those involved in the expansion of the Flatpack Democracy model who have informed the IfF+ story outlined in this book in the context of climate and ecological emergencies. A special thanks to Peter Macfadyen who always made himself available for fascinating catch-ups on developments in the New Economy and Flatpack Democracy movements.

The publishers have also been incredibly patient with me as I have waited for various ducks to align so that I can finish my manuscript, including taking stock of the pandemic that was nowhere on the horizon when I conducted my initial empirical research and other recent administrative changes in Somerset and beyond. While I have endeavoured to ensure that what is published has been appropriately fact-checked, if anything has crept in that is disputed then I apologise. Please get in touch to let me know if anything needs to be corrected in any later iteration of this work. For the sake of readability, acronyms are reintroduced in each chapter (considering that some chapters may be read online or in isolation). I hope this doesn't seem too repetitive for those of you who read this book in full.

I'd like to express thanks to the Economic and Social Research Council (ESRC) for recognising the value of my work and funding a one-year post-doctoral fellowship, based at the University of Surrey's Centre for Environment and Sustainability (CES) and co-hosted within the Centre for the Understanding Sustainable Prosperity (CUSP), which has a fabulous cohort of researchers and practitioners advancing cutting-edge research on sustainable prosperity and how to design an inclusive future. I am also grateful to all my colleagues at the Middlesex University's Centre for Enterprise, Environment and Development Research (CEEDR) for surrounding me with inspiring ideas and fascinating research projects, from which I learn a lot daily and the insights I have gained from working at CEEDR have strengthened the overall insights that have further refined the content of this book. Special thanks to Prof John Sturzaker for responding to a last-minute fact-check request and Prof Nick Gallent for giving me some writing tips on how to stop procrastinating and write with a sustained effort after having stopped-started in my attempts to finish this book, waiting for a block of time to write that never comes. Thankfully, there is now a tangible output.

Praise for *The Politics of Transition*

"A timely exploration of the politics of placemaking and the limits of localism in meeting radical ambitions for community-led resilience. It offers important insights to communities and councils wanting to generate novel ideas and challenge the status quo, based on a balanced account of the conditions that led to Frome becoming a hotbed of ideas aiming to realise the New Economy."

—Peter Macfadyen, instigator of Sustainable Frome and the Flatpack Democracy movement, author of *Flatpack Democracy: a DIY Guide to Creating Independent Politics and Flatpack Democracy 2.0: Power Tools for Reclaiming Local Politics*, and former Leader of Frome Town Council

Contents

Abbreviations

25-YEP	25-Year Environment Plan
AGM	Annual General Meeting
BACCC	Bristol Advisory Committee on Climate Change
BIES	Department of Business, Industry, Energy and Strategy
BNG	Biodiversity Net Gain
BREEAM	Building Research Establishment Environmental Assessment Method
CO_2	Carbon Dioxide
CCC	Committee on Climate Change
CEE	Climate and Ecological Emergency
CEEDR	Centre for Enterprise, Environment and Development Research
CES	Centre for Environment and Sustainability
CfSH	Code for Sustainable Homes
CIC	Community Interest Company
CIL	Community Infrastructure Levy
CLT	Community Land Trust
CoP	Conference of the Parties
COVID-19	Coronavirus disease 2019
CRtBO	Community Right to Build Order
CSE	Centre for Sustainable Energy
CUSP	Centre for the Understanding of Sustainable Prosperity
DCLG	Department of Communities and Local Government
DECC	Department for Energy and Climate Change
DfL	'Down from London'
DLUHC	Department for Levelling Up, Housing and Communities
EIP	Environmental Improvement Plan
ELMS	Environmental Land Management Scheme

EPC	Energy Performance Certificate
EU	European Union
FFCC	Food Farming and Countryside Commission
FFN	Frome Neighbourhood Network
FiT	Feed-in-Tariff
FoE	Friends of the Earth
FRECo	Frome Renewable Energy Co-op
FTC	Frome Town Council
GDPR	General Data Protection Regulations
GFI	Green Finance Institute
GI	Green Infrastructure
HCA	Homes and Communities Agency
HSR	Housing Standards Review
IfF	Independents for Frome
ISSB	International Sustainability Standards Board
IUC	International Urban Cooperation
LCN	Local Community Network
LDF	Local Development Framework
LEP	Local Enterprise Partnership
LGA	Local Government Association
LGS	Local Green Spaces
LNPs	Local Nature Partnerships
LPA	Local Planning Authorities
MAP	Multi-Actor Perspective
MCTI	Market and Coastal Towns Initiative
MDC	Mendip District Council
MHCLG	Ministry of Housing, Communities and Local Government
MLP	Multi-Level Perspective
MoU	Memorandum of Understanding
MP	Member of Parliament
$MtCo_2e$	Million tonnes of carbon dioxide equivalents
NA	Neighbourhood Area
NALC	National Association of Local Councils
NAO	National Audit Office
NDCs	National Determined Contributions
NDO	Neighbourhood Development Order
NDP	Neighbourhood Development Plan
NGO	Non-Governmental Organisation
NHS	National Health Service
NIMBY	Not In My Back Yard
NPPF	National Planning Policy Framework
OEP	Office of Environmental Protection

OFSTED	Office for Standards in Education, Children's Services and Skills
OIF	Outcome Indicator Framework
ONS	Office of National Statistics
OPL	One Planet Living
PCPA	Planning and Compulsory Purchase Act
PD	Permitted Development
PES	Payments for Environmental Services
Ph.D.	Doctor of Philosophy
QCA	Qualitative Content Analysis
R2o	Route to Zero
RDAs	Regional Development Agencies
RICS	Royal Institution of Chartered Surveyors
RQ	Research Question
SCC	Somerset County Council
SDGs	Sustainable Development Goals
SDR	Sustainability Disclosure Standards
SEA	Strategic Environmental Assessment
SMEs	Small and Medium-Sized Enterprises
SNA	Social Network Analysis
SNM	Strategic Niche Management
SPDs	Supplementary Planning Documents
SWIMBY	Something Wonderful in my Back Yard
TCFD	Taskforce for Climate-related Financial Disclosures
TCPA	Town and Country Planning Association
TDS	Town Design Statement
TI	Transition Initiative
TN	Transition Network
TNFD	Taskforce for Nature-related Financial Disclosures
ToR	Terms of Reference
UK	United Kingdom
UoR	University of Reading
US	United States
V4F	Vision4Frome
WoW	Ways of Working
XR	Extinction Rebellion
ZCH	Zero-Carbon Homes

List of Figures

List of Tables

CHAPTER 1

Introduction: Political Fragmentation and Opportunity for Climate Action in a Changing World

Navigating the Climate Challenge: Moving Beyond Business-as-Usual

We are living in an age of accelerating awareness of our need to reduce our environmental impact. If the news headlines, multiple reports and academic articles are to go by, to put it lightly, as a species and with us the many flora and fauna upon which our actions are inextricably tied, we are in trouble. The current rate of environmental destruction, species decline, population growth, the use of fossil fuels, pollution, declining food security, drought, flooding and obesity are all trends that, left unchecked, will adversely impact human and environmental well-being.

The book is set within growing attention to the 'political' within sustainability transitions—the extent to which societies are moving towards more sustainable forms of living and where social and technical re-orderings result in sustainable goal-directed pursuits (Schot & Geels, 2008; Kemp et al., 1998). The need to radically transform our economy and relationship to nature has led some commentators to suggest that 'globalisation is over', or at least on its descent (Larry, 2015). If true, this could also bring with it a challenge to liberal democracy in the evolution of planet-positive (democratic) systems. As the political theorist Francis Fukuyama (1992) predicted, democracy is in crisis in many parts of the world, reflected in the rise of the right-wing government and the ascent of war in Europe with Russia's

A. Burnett, *The Politics of Transition*, Palgrave Studies in Sub-National Governance,
https://doi.org/10.1007/978-3-031-40769-7_1

invasion of Ukraine. There are also promising signs of political experimentation that might realise democracy exercised within planetary boundaries, the politics of which is the subject of the chapters that follow.

Times of crisis create opportunities for a shift in how we perceive ourselves and our relationships with others and to image alternatives (Jackson, 2009; Escobar, 2015). At different scales, there is an increasing convergence of frameworks and institutional architecture to respond to climate change, such as the United Nation's Sustainable Development Goals (SDGs). There is now widespread global recognition of the need to act on climate change. Even some countries previously resistant to climate action have followed suit, such as China and the United States (US) now committing to net-zero 2060 and 2050 targets. The Biden Administration has secured large-scale investment in decarbonised energy through the 2022 Inflation Reduction Act. Europe's European Green Deal aims to achieve carbon neutrality by 2050. Such ambitions are also reflected in the Club of Rome's planetary emergency plan, which sets out the foundations for long-term economic and societal shifts, underpinned by human and planetary prosperity and well-being.

Awareness of environmental issues is on the rise. According to a YouGov survey in May 2019 (YouGov, 2019), climate change was the third most pressing issue for respondents in the United Kingdom (UK). A quarter (27%) of Britons cited the environment as their top three issues facing the country, putting it behind only Brexit (67%) and health (32%). The awareness of climate change is particularly high among 18- to 24-year olds, rising to 45%, making it their second biggest concern behind Brexit (57%). This surge was also mirrored by the widespread direct action by Extinction Rebellion (XR) and Greta Thunberg in 2019—who appealed to much of the global youth population. These events, together with the Conference of the Parties in Paris in 2019, stimulated a cascade of climate (and a lesser amount of climate and ecological) emergency declarations. In May 2019, the UK government also declared a climate emergency.

The recent war in Ukraine is having detrimental ripple effects across the globe, ranging from food shortages to oil price hikes and energy shortages. Inflation is rising, leading to a cost-of-living crisis. The political upheaval of Brexit and the war in Ukraine will change the UK's political and economic structures within Europe and globally. These trends will likely take their toll on having the means to deliver on climate action as households, businesses and government seek to rebalance priorities in the wake of the fiscal tightening. Globalisation has arguably left many scars on nature and our political and economic systems.

A 2021 Accenture report which mapped the impacts of the COVID-19 pandemic on SDG targets and indicators concluded that the pandemic posed severe near-term challenges and hinders longer-term progress against all 17 goals (O'Reilly et al., 2021). In June 2020, the Committee on Climate Change (CCC, 2020a) published its annual Progress Report to Parliament, which assesses UK's progress in reducing greenhouse gas emissions. It recommended that the UK Government "seize the opportunity to turn the COVID-19 crisis into a defining moment in the fight against climate change" and that more action needed to happen in the following areas (see CCC, 2020b):

- Low-carbon retrofits and buildings that are fit for the future.
- Tree planting, peatland restoration and green infrastructure.
- Energy networks must be strengthened.
- Infrastructure to make it easy for people to walk, cycle and work remotely.
- Moving towards a circular economy, for example, reducing waste, upcycling, sharing and local production.
- Investing in the UK's workforce, lower-carbon behaviours and innovation (e.g., reskilling and retraining programmes, encouraging positive behaviours, innovation funding).

How might one begin to demonstrate how these shifts take place, which actors and institutional shifts have caused it, and what role might planning play in encouraging a net-zero society? What role do our local institutions play in upholding neoliberal practices or preventing development alternatives from breaking through?

How much and what type of growth has also been questioned by those pursuing a degrowth agenda (Escobar, 2015) or the need for a different kind of socio-economic framework in the context of a green recovery and the urgent need to address the climate and ecological crisis (UKGBC, 2021). Recent UK policy has actively sought to remodel the relationship of key sectors, such as agriculture and the planning process, to nature and the environment in legislation such as the Agriculture Act 2020 and the Environment Act 2021. Yet what constitutes sustainable development is something of a hot topic. There are often different 'shades' of what is meant by being 'green' (Muir et al., 2000; Dryzek, 2005) yet sustainable development often mirrors the widely cited definition in the Brundtland

Report of economic, social and environmental pillars of sustainability (The Brundtland Commission, 1987).

There is potential for new political identities to be forged and with those new structures and institutions to be created (although it is recognised that new is not always progressive). This is a book about the potential to transform our political landscape, and through this, our relationships with each other through placemaking—the art of making places—and the role of planning towards these ends. For instance, how can we design systems that are inclusive and sensitive to designing resilient and empowered communities, at scale?

As well as probing these issues through a discussion of UK environmental and planning policy and approaches, we home into these by examining a particular case of Frome, an English market town in the county of Somerset that is experimenting with a new kind of ultra-participatory, inclusive, socially and environmentally orientated politics that eschews traditional party politics.

Indeed, it appears that a new wave of eco-political discourse is gaining traction, emphasising notions of degrowth and socio-cultural experimentation, including a movement towards cultivating a New Economy based on collaborative and progressive governance and culture. This highlights what Purcell (2009, p. 144) terms the "cracks and instabilities" within neoliberal modes of governance.

Peer-to-peer technologies, such as crowdfunding and internet-based movements are helping to link disparate actors through networked platforms, which could give rise to more rapid diffusion and uptake of new development models. For instance, the Global Covenant of Mayor's initiative and related working platforms are encouraging global networks to promote local government action on climate change and peer support to achieve these goals.[1] In the European Union (EU), cities are also paired with cities from regions facing related sustainable development challenges; here they are encouraged to share knowledge and best practice on sustainable urban solutions and the development of local action plans through a knowledge-exchange programme, the International Urban Cooperation (IUC). For others, grassroots players are seeking in some cases to circumvent mainstream players, such as corporations and government, and to develop radical solutions to common issues such as housing, energy and

[1] https://www.globalcovenantofmayors.org/

nature recovery through community-led governance such as community energy and Community Land Trusts (CLTs).

Some have suggested that these trends reflect a 'new era' in politics, where the innovative and the everyday are central as new actors seek to destabilise the status quo (Adnan, 2016) as an effect of globalisation and technology (Castells, 2007). Multi-stakeholder processes can encourage awareness of different perspectives and can engender innovation through improved interaction, investment, joint resource mobilisation and policy advocacy.

These recent tendencies also invite further debate within development and planning institutions on the relative importance of climate change and the environment over profit and economic growth. Meanwhile, recent UK government austerity measures may have begun to open up opportunities for social enterprises to fill the spaces between market and state. Advances in technology have increased the affordability and availability of renewable energy (IRENA, 2017). Narratives of 'Build Back Better' permeated recent political and social discourse, which call for a more resilient and nurturing society through shared values of well-being with opportunities to reform democracy, tackle inequality and promote cohesive communities.

Of particular importance is how climate emergencies can address the issue of the UK's energy-inefficient housing stocks, some of the worst in Europe. As of 2022, emissions from housing are estimated to be 17% of the UK's total carbon emissions (DESNZ/ONS, 2023). The National Housing Federation estimates that England's 25 million (m) homes account for the equivalent carbon dioxide (CO_2) emissions of 28m cars (equating to 58.5 tonnes of CO_2 a year) (NHF, 2021). The Committee on Climate Change (CCC) has also suggested that while the UK's response to climate change has made inroads in the energy sector but not many gains in housing (CCC, 2020b). Transport and energy still account for the main sources of carbon emissions (DESNZ, 2023).

Which discourses of 'sustainable development' are invoked, and by whom, can either seek a more transformative agenda or encourage a post-political 'fix' in development and planning practice (Allmendinger & Haughton, 2012; North, 2011). As Tompt suggests, "the models are there. How can we create the conditions to put them into practice?" (Tompt, 2016, p. 11). Meanwhile, as Padgett and Powell (2012, p. 1) suggest in their chapter "The Problem of Emergence", theoretical approaches can often suggest how or what alternatives exist, but these

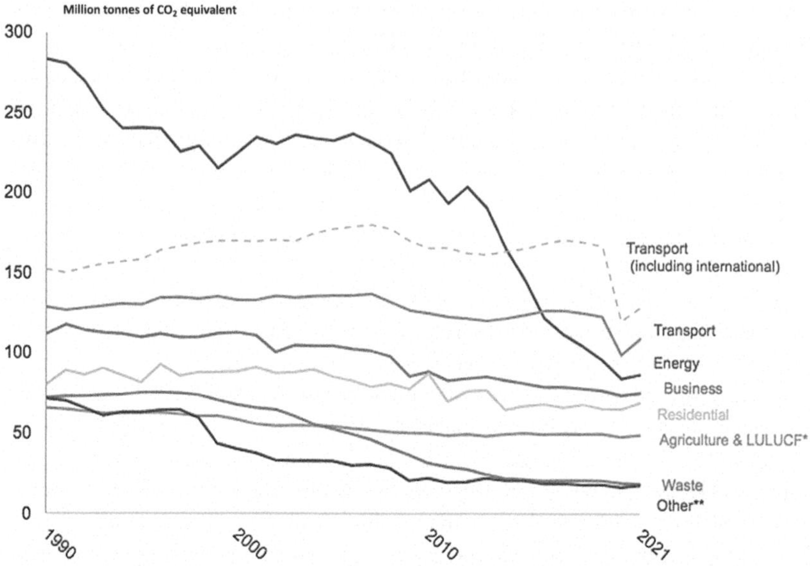

Fig. 1.1 Greenhouse gas emissions by sector, 2021 (DESNZ, 2023)

"have little to say about the invention of new alternatives in the first place". Therefore, innovative responses require an in-depth understanding of how these ideas carry traction.

Objectives of This Book

This book seeks to understand the extent of these cracks and instabilities at a local level and how grassroots and lower tiers of government might seize them and what are the most appropriate arenas to cultivate new imaginaries of development in a more 'sustainable' future. Here, we reflect on how innovative, and potentially transformative, ideas are carried within and between government, the implications for upscaling and increasing the resilience of community initiatives in different settings, and the role of market towns (a town of moderate size where a regular market is held, a term which has roots in the Middle Ages) as settings to promote sustainability transitions and their replicability to different areas, including more rural settings.

In response to Tompt's question and Padgett and Powell's challenge, I ask in what contexts can 'New Economy' sustainability innovations work and why, and what implications does this have for the scaling up of innovations for climate and ecological emergencies? For instance, unpacking how actors come together, through which formal and informal processes, and what motivations and shared interests and experiences enabled ideas to emerge from a latent to an active state to challenge growth-centred logics. This book is a quest to investigate the scope for implementing innovative, 'sustainable' development alternatives through mainstream institutions or initiatives at the local scale, namely at the town and parish council level.

Local-level engagement with the planning process is key to ensuring decisions relating to place and its function and identities reflect local interests and promote a sustainable approach to land management in built and natural environments (Williams, 2013, Carroli, 2018; Smedby & Quitzau, 2016). How new forms of environmental governance and their measurement, such as biodiversity net gain (BNG), which seeks a minimum 10% uplift in biodiversity arising from new developments—enshrined in the Environment Act 2021—are integrated into existing environmental place-making networks is also a timely and important issue to understand their transformative potential and their relevance to community-based agendas and networks. Local nature recovery strategies (also embedded in the Environment Act) also offer opportunities for community participation in hybrid public, private and civil society partnerships. Rapidly expanding new markets in natural capital, where private investors are looking to pay landowners, farmers and potentially communities to improve nature-based outcomes also offer new ways to resource environmental action at the grassroots, though these markets also have potential risks if contracts do not encourage fair outcomes. New environmental metrics launched in early 2023 in the Outcomes Indicator Framework, tied to the Environmental Improvement Plan and the 25-Year Environment Plan could also, potentially, play a transformative role in encouraging coherent responses by and with government at different scales.

The following chapters dissect what transition concepts mean when protagonist actors—many rooted in non-partisan backgrounds—occupy formal political roles and the barriers to entry (and opportunities) in using the state to challenge traditional conceptions of politics and development. Through a multi-level account of transition dynamics (that is focusing on the actors and entities involved in seeking to influence practices, behaviours, policies and plans to achieve sustainable development at different scales), this book examines what happens when grassroots environmental organisations engage with more institutional actors or willingly co-opt themselves within the systems they seek to change.

Neighbourhood plans are one potential 'arena' in which to explore these issues. These plans are community-led planning instruments created under the 2011 English Localism Act and are also known as Neighbourhood Development Plans (NDPs). This reflects their status as being part of a local planning authority's planning framework, once 'made' (adopted) after passing a local referendum). Initially, this study was a focus on 'low-carbon' neighbourhood plans, where I was interested in how they were being used by communities to channel sustainability ambitions, particularly more radical ideas that departed from business-as-usual approaches to planning and development (i.e., larger housing developments divorced from community input or a sense of place, or where profit is prioritised over impacts on the environment).

However, as I progressed my research on this topic, interviewees increasingly began to highlight key issues relating to place identity—its construction, politicisation and institutionalisation—revealing the politics of transition in cultivating innovative placemaking and alternative development models. Identity can help to encompass the psychosocial qualities of identity in-use, including the propensity for 'othering' of those who are not considered part of the 'in-group'—groups to which individuals and groups feel they belong. What is interesting is whether 'niche' protagonists—be they local political actors or social movement members participating in formal arenas such as neighbourhood plans—possess and retain these identities if they occupy formal spaces of power (such as by becoming councillors themselves), and what role different 'regimes' have on cultivating and contesting development alternatives (such as land-use planning or party-political systems).

This book also responds to a call for research on whether green policies and initiatives might be more readily accepted under the guise of new forms of political action, such as independent politics (Macfadyen, 2014). For the sake of offering a definition to frame these discussions, I put forward that 'sustainability' is the continued circulation of positively reinforcing socially equitable, environmentally regenerative and economically inclusive outcomes that do not stretch the earth's planetary limits. I strongly believe these reinforcing relationships can be cultivated through a mechanism called regenerative value, which can be done at a societal level and catered to different sectors (see Burnett, 2022a, 2022b).

We take a socio-institutional account of sustainability transitions (Loorbach et al., 2017) to understand whether *Transition* has influenced the conditions for development alternatives to take root and assess the role of planning and placemaking dynamics to shape resultant transition pathways. In this book, we will assess the role of planning within sustainability transitions and unpack power, politics and institutional arrangements, role boundaries and transfer on transition processes, and place and scale in transition. Such an approach is critical to understand the drivers for such a change or whether a niche movement for change can penetrate planning and its related socio-technical system, the development sector.

Key concepts are useful to help understand the contribution of planning to broader transition processes, for instance, the 'niche' (i.e., radical protagonists) and the 'regime' (i.e., institutions and state rules) and how they move in *transition arenas*. That is, a term used by sustainability transition scholars, notably those exploring Transition Management (Loorbach & Rotmans, 2010), to describe the purposeful governance arrangements to solicit 'transitions'. It is recognised here that these may not always be purposeful, and some arenas may be subject to group capture for subversive (i.e., niche-directed) transitions.

Whether such ideas remain 'niche' or if these can penetrate planning policy better if protagonists occupy positions of the state is a core objective of this book. To do so, we explore the interplay between local administrative political structures and their relationship in bridging the niche and regime through spatially situated networks in Frome. Changes to the occupation of political roles on local placemaking networks can give a longitudinal and multi-level perspective on placemaking transitions, if any. This raises a question over how development alternatives travel from the grassroots (placemaking networks) through planning and how these spaces might act as 'terrains of resistance' for imagining alternative futures and degrees of their institutionalisation, or not, into mainstream processes of governance and planning arenas (see Routledge, 1996).

Attention to the political-planning praxis and how this affects the application of localism is an important consideration to the institutional arrangements for niche innovations in sustainability transitions. Particularly where these can highlight: i) the constraining and enabling functions of systems and their capacity for either steer and emergence (Nykvist & Turnheim, 2015; Bergek et al., 2015; Avelino & Wittmayer, 2015); ii) how interaction between local and higher tiers of government can foster transition pathways (see Späth & Rohracher, 2012); iii) mapping 'political fault lines' within and outside transitions arenas, and iv) whether this can highlight fundamental tensions and opportunities (Kenis et al., 2015).

Meanwhile, planning theory has not yet fully addressed the growing, interdisciplinary body of research exploring how to cultivate sustainability transitions. There remains a need to understand the broader role of placemaking—a collective effort of those within a given spatial remit to reimagine or stabilise local character and form—and a historical understanding of social network influence on the nature of particular constellations of 'transition' and the enactment of roles that promote, or frustrate, sustainability transitions. Placemaking itself has morphed from narrow design strategies towards a tool for empowerment by change agents (see Strydom et al., 2018); the term's relation to planning renders the interaction of 'community-based' placemaking (which may be considered more 'niche') and spatial planning (often seen as more 'regime') a fruitful study to explore the praxis between informal and formalised transition strategies and arenas.

The chapters that follow showcase a behind-the-scenes, critical analysis of the strategies and politics advancing sustainability in a context of 'progressive' and potentially transformative local politics. We will focus on planning's contribution to sustainable development alternatives at a local level within the English planning policy context and what linkages 'planning arenas', such as NDPs, have with broader placemaking and political arrangements. Notably, whether land-use and urban planning tools and practice can contribute to socio-technical, and socio-political transitions, and the development of a wider body of planning transitions theory (see for instance, Lukkarinen et al., 2023); that is, how can planning solicit environmental, social or even political sustainability transitions and what these concepts mean within a given spatial context.

Frome: An Exemplar Case Study of Localised Environmental (Political) Action

We home in on an innovative case study of the market town of Frome, which is becoming increasingly known for its bold green action and its fiercely 'independent' character. Frome Town Council is, since 2011, controlled by a group of local people under the guise of the 'Independents for Frome (IfF)' group. IfF is a non-party-political organisation and has, by bypassing party politics, been able to achieve many positive initiatives by breaking the mould of adversarial, often divisive party politics, including reconstructing the administration of the town council towards non-partisan Ways of Working (WoW).[2] Many initiatives under IfF, such as seeking to bestow 'nature rights' on the local River Frome, are what some scholars and activists term the 'Commons'—the community management of resources with societal and environmental benefits.

This in-depth research on placemaking in Frome homes in on the role of social movements in cultivating community-led, development alternatives within and around mainstream political and planning 'regimes'. The findings hope to contribute to wider (popular) debates and movements, including transition, placemaking, local politics and planning, by revealing the ingredients of successful strategies (or otherwise) of fringe ideas becoming mainstream and institutionalised in and around the fabric of government institutions.

By showcasing development and governance alternatives in Frome, this book seeks to shed light on the complexities and realities of Frome's success, either for those seeking to replicate the Flatpack Democracy model (Macfadyen, 2014) or for those involved in community-based planning (e.g., neighbourhood planning) who seek to cultivate positive placemaking. Frome's experiences in using the planning process as a vehicle to secure strong environmental and social benefits at the local level and the reasons for success, or otherwise, can provide insights in how to capture and reorientate local institutions towards these ends.

Frome's story also reveals the politics and tensions of 'independent' actors challenging established institutions, particularly when using rights that promised to deliver power to ordinary people under the government's UK's localism policy agenda, which claims to "give power" to local-level actors.

FTC also used its NDP to promote strongly relocalised ideas and practices on cohousing, self-build, shared space schemes and One Planet Living (OPL) principles, which was 'made' (adopted) in October 2016. By taking a deep dive into the mechanics of this process, we assess the extent to which Neighbourhood Plans strongly filtered the amount of power available at the lowest tiers of government or entrenched local networks. Meanwhile, concrete examples of the politics of transition at a local level can point to the conditions for effective change and the limits to

[2] See Appendix C—IfF's Ways of Working.

community influence in spaces that, in theory, seek to promote community involvement.

Frome's case builds on Bulkeley's (2019) challenge to critically engage with the types of knowledge being produced on climate change and their unfolding processes and dynamics (see Avelino, 2017). Two 'tribes' of the New Economy (see Real Economy Lab, 2016)—(i) *Transition* and relocalisation and (ii) government and politics—are explored in detail as potential transition arenas to understand their degrees of intersection as conditions for (re-localised) sustainability transitions. As such, in the chapters that follow, we will explore the fuzziness and clearly defined remit of both planning and political 'regimes' (i.e., defined institutional boundaries; see Nill & Kemp, 2009), the extent to which these different facets of (local) government were attuned to more radical or 'niche' sustainability agendas, and the politics within such arenas.

Specifically, key entities of relocalisation, including local Transition Initiatives (part of the Transition town movement) and their related spin-off groups, are mapped in relation to their ties to mainstream development structures, notably the NDP process and Frome Town Council (FTC). Meanwhile, exploring the duality of the 'niche' and the 'regime' within placemaking networks, that is, how much grassroots or formal organisations can be said to have either of these characteristics, within formalised and informal structures provides an extended view of the remit of planning on placemaking to solicit conditions for (relocalised) sustainable development.

By exploring different actors, their relationships, agendas and engagement in different 'arenas' of placemaking, this book seeks to explore:

- How transition arenas contain and mediate power relations (i.e., actors enrolled or excluded from such processes).
- The influence of social and spatially rooted ties on the governance of these arenas.
- The relationship between governance, culture, personality and autonomy, and ways in which these factors affect the strategies to provoke transitions in different settings.
- Whether transitions can be planned and the balance between steer and emergence in cultivating development alternatives.

While many interviews carried out for this book were carried out in 2015, I have continued to pay a keen eye to developments in Frome in terms of the relocalised agenda and its wider relationship to social movement 'New Economy' dynamics. As such, the findings deriving from the interviews have been supplemented with additional secondary research on 'sustainable development'-related outputs in Frome and the wider ripple effects to other areas (and vice versa) through my continued engagement with those I interviewed some five or six years ago. I have also complemented this work with more recent research on the spread of Flatpack Democracy to other areas and the monitoring and evaluation (or not) of climate action and desirable features of such a process for different stakeholders. The timing of the interviews, rather than making this data irrelevant, is even more pertinent if looked at from a longitudinal perspective (i.e., looking at events at different points in time), because the impact of sustainable placemaking action and the relevance of Frome's NDP can be situated within the wider context of planning for transitions in the context of climate and ecological emergencies.

The next chapter (Chap. 2) summarises the overarching environmental policy framework in the UK and the role of local government in achieving sustainable development, including outlining key roles and responsibilities of local government and their capacity to generate innovation in this area through formal and informal powers.

Chapter 3 explores the role of the planning system in enabling, or preventing, innovation in sustainability within the context of the English National Planning Policy Framework (the NPPF). Here we also takes deeper look at the potential for environmental and political movements to engender development alternatives and community-led, relocalised placemaking.

Chapter 4 details why Frome is of particular relevance, the town's characteristics, the key actors involved and the role of spatial affect and personal ties in shaping the character of placemaking networks at the local level and how this affected IfF's route to power.

Chapter 5 assesses the experiences of those involved in using Frome's Neighbourhood Plan in order to realise these goals to advance relocalised development.

In Chap. 6, we explore the various conditions that enabled relocalised political innovation in Frome and how specific relationships helped to give rise to emergent conditions that allowed IfF and to consolidate power and pursue particular sustainability ambitions and development alternatives in the town.

Chapter 7 outlines Frome's development challenges and the relationship with relocalisation, planning and the development industry as well as

Frome's potential response to these challenges through the anticipated benefits of the localism agenda to deliver locally led solutions (and the role of the NDP process in actualising mechanisms to promote such an identity).

Chapter 8 turns to examining whether localism can indeed be used for purposes "not quite as it intended" and what the conditions of using localism effectively are, particularly for more 'subversive' political and placemaking objectives.

In Chap. 9, we explore how the informal relational ties that were essential in building Sustainable Frome played out not only in more formal governance of engaging in the neighbourhood plan but also in influencing wider issues of apathy and civic engagement at a broader scale.

Chapter 10 assesses what are the consequences of the occupation of the state by protagonists in the form of niche-directed transitions and the intended and unintended effects on local placemaking dynamics (notably Sustainable Frome) or between tiers of government and broader (non-party) political agendas.

In Chap. 11 we assess how IfF's independent political agenda had already begun to deeply affect not only the politics of the town council but its relationship with other tiers of government. We will probe further into the politics of claiming 'independence' in local elections and as the Flatpack Democracy movement sought to scale up and out in the context of wider administrative reform of within local government in England.

We conclude by reflecting on the potential for the Flatpack Democracy movement to scale up and out and the repercussions of the COVID-19 pandemic on placemaking and innovative political experimentation. Here, I reflect on the need to create, capture and circulate sustainable values that underpin the efforts of us all as we seek to find workable solutions to the issues and opportunities we collectively face, in the UK and globally.

References

Adnan, I. (2016). *21C politics: Is the party over or is it just beginning?* Compass.

Allmendinger, P., & Haughton, G. (2012). Post-political spatial planning in England: A crisis of consensus? *Transactions, 37*(1), 89–103.

Avelino, F. (2017). Power in Sustainability Transitions: Analysing power and (dis) empowerment in transformative change towards sustainability *Abstract Environmental Policy and Governance, 27*(6) 505–520. https://doi.org/10.1002/eet.v27.6 https://doi.org/10.1002/eet.1777

Avelino, F., & Wittmayer, J. M. (2015). Shifting power relations in sustainability transitions: A multi-actor perspective. *Journal of Environmental Policy & Planning, 18*(5), 62–649.

Bergek, A., Hekkert, M., Jacobsson, S., Markard, J., Sandén, B., & Truffer, B. (2015). Technological innovation systems in contexts: Conceptualizing contextual structures and interaction dynamics. *Environmental Innovation and Societal Transitions, 16*, 51–64. https://doi.org/10.1016/j.eist.2015.07.003

Bulkeley, H. (2019). Navigating climate's human geographies: Exploring the whereabouts of climate politics. *Dialogues in Human Geography, 9*(1), 3–17.

Burnett, A. (2022a). Securing a regenerative and just planning system. *Town & Country Planning, 91*(2), 123–128.

Burnett, A. (2022b). Operationalising performance measurement and incentives for 'regenerative value' to improve net sustainability outputs. In G. Vélez-Cuartas & O. Romero-Goyeneche (Eds.), *Transformative metrics: Contributions to the studies for monitoring and evaluating how science, technology, and innovation can address social and environmental challenges.* University of Antioquia. https://bibliotecadigital.udea.edu.co/bitstream/10495/31313/1/VelezGabriel_2022TransformativeMetricsContribitions.pdf

Carroli, L. (2018). Planning roles in infrastructure system transitions: A review of research bridging socio-technical transitions and planning. *Environmental Innovation and Societal Transitions, 29*, 81–89.

Castells, M. (2007). Communication, power and counter-power in the network society. *International Journal of Communication, 1*, 238–266.

CCC. (2020a). *2020 progress report to Parliament.* Climate Change Committee.

CCC. (2020b). *Letter: Building a resilient recovery from the COVID-19 crisis to prime minister Boris Johnson.* Climate Change Committee.

DESNZ. (2023). 2022 UK greenhouse gas emissions, provisional figures. Department for Energy Security and Net Zero/National Statistics. Retrieved from https://assets.publishing.service.gov.uk/government/uploads/system/uploads/attachment_data/file/1147372/2022_Provisional_emissions_statistics_report.pdf

Dryzek, J. S. (2005). *The politics of the Earth.* Oxford University Press.

Escobar, A. (2015). Degrowth, post-development, and transitions: A preliminary conversation. *Sustainability Science, 10*(3), 451–462.

Fukuyama, F. (1992). *The end of history and the last man.* Free Press.

IRENA. (2017). *Renewable power generation costs in 2017. Key findings and executive summary.* .

Jackson, T. (2009). *Prosperity without growth? The transition to a sustainable economy.* .

Kemp, R., et al. (1998). Regime shifts to sustainability through processes of niche formation: The approach of strategic niche management. *Technology Analysis and Strategic Management, 10*, 175–195.

Kenis, A., et al. (2015). Unravelling the (post-)political in transition management: Interrogating pathways towards sustainable change. *Journal of Environmental Policy & Planning, 18*, 568–584.

Lab, R. E. (2016). Real Economy Lab. http://reconomy.org/ [careful of malware on this site]

Larry, E. (2015). Does Doha trade talks' failure suggest second age of globalisation is over? *The Guardian*. Guardian News & Media Limited.

Loorbach, D., Frantzeskaki, N., & Avelino, F. (2017). Sustainability transitions research: Transforming Science and practice for societal change. *Annual Review of Environment and Resources, 42*, 599–626.

Loorbach, D., & Rotmans, J. (2010). The practice of transition management: example and lessons from four distinct cases. *Futures, 42*, 237–246. https://doi.org/10.1016/j.futures.2009.11.009

Lukkarinen, J. P., Nieminen, H., & Lazarevic, D. (2023) Transitions in planning: transformative policy visions of the circular economy and blue bioeconomy meet planning practice. *European Planning Studies 31*(1) 55–75. https://doi.org/10.1080/09654313.2022.2060706

Macfadyen, P. (2014). *Flatpack democracy: A DIY guide to creating independent politics.* Eco-Logic Books.

Muir, K., et al. (2000). Shades of green in local authority policy-making: A regional case study. *Area, 32*(4), 369–382.

NHF. (2021). England's leaky homes are a greater threat to climate than all of its cars. Retrieved June 4, 2023, from https://www.housing.org.uk/news-and-blogs/news/englands-leaky-homes-greater-threat-to-climate-than-cars/

Nill, J., & Kemp, R. (2009). Evolutionary approaches for sustainable policies: From niche to paradigm? *Research Policy, 38*, 668–680.

North, P. (2011). The politics of climate activism in the UK: A social movement analysis. *Environment and Planning A, 43*, 1581–1598.

Nykvist, B., & Turnheim, B. (2015). Deliverable D4. 3 Report reviewing the literature on policy and governance of sustainability transitions. Transition Pathways to Sustainable Low-Carbon Societies: EU-funded project grant agreement no: 603942.

O'Reilly, K., et al. (2021). *Delivering on the promise of sustainability.* Accenture.

Padgett, J. F., & Powell, W. (2012). *The emergence of organizations and markets.* Princeton University Press.

Purcell, M. (2009). Resisting neoliberalization: Communicative planning or counter-hegemonic movements? *Planning Theory, 8*(2), 140–165.

Routledge, P. (1996). Critical geopolitics and terrains of resistance. *Political Geography 15*(6–7) 509–531. https://doi.org/10.1016/0962-6298(96)00029-7

Schot, J., & Geels, F. W. (2008). Strategic niche management and sustainable innovation journeys: Theory, findings, research agenda, and policy. *Technology Analysis and Strategic Management, 20*(5), 537–554.

Smedby, N., & Quitzau, M. B. (2016). Municipal governance and sustainability: The role of local governments in promoting transitions: Municipal governance and sustainability. *Environmental Policy and Governance, 26*(5), 323–336.
Späth, P., & Rohracher, H. (2012). Local demonstrations for global transitions—Dynamics across governance levels fostering socio-technical regime change towards sustainability. *European Planning Studies, 20*(3), 461–479.
Strydom, W., et al. (2018). Exploring theoretical trends in placemaking: Towards new perspectives in spatial planning. *Journal of Place Management and Development, 11*(2), 162.
The Brundtland Commission. (1987). *Our common future*. Oxford University Press.
Tompt, J. (2016). *The meaning of Brexit? It's the movement. STIR to Action*. Dorset, STIR.
UKGBC. (2021). *Building a just transition to net zero*. UKGBC.
Williams, J. (2013). The role of planning in delivering low-carbon urban infrastructure. *Environment and Planning B: Planning and Design, 40*, 683–706.
YouGov. (2019). *Concern for the environment at record highs*. Retrieved September, 9, 2020, from https://yougov.co.uk/topics/politics/articles-reports/2019/06/05/concern-environment-record-highs

CHAPTER 2

Sustainability Transitions in the Context of the English Localism Regime

Discourses of Transition

Which discourses of 'sustainable development' and climate change are used, and by whom, can either advance a transformative agenda or encourage a post-political 'fix' in development and planning practice (Allmendinger & Haughton, 2012). At the same time, multi-stakeholder processes—such as citizen assemblies and community-led planning—can help improve capacities, identify constraints and opportunities and support wider resource mobilisation to achieve shared goals through processes of constructive negotiation (Leeuwis, 2000).

The field of sustainability studies known as sustainability transitions explores the normative (i.e., purposeful and value-driven) quest for different actors to navigate existing institutions and 'regime' institutions to deliver innovative and impactful sustainable outcomes. For instance:

- Changing governance to solicit change (i.e., 'transition management' through specific 'transition arenas').
- Policy-led change and its wider structuring effects on society.
- Behaviour change and psycho-social approaches.
- The politics of seeking to achieve these changes through multi-stakeholder networks.

A. Burnett, *The Politics of Transition*, Palgrave Studies in Sub-National Governance,
https://doi.org/10.1007/978-3-031-40769-7_2

Within this body of literature, researchers often use the terms transition and transformation to describe the longevity and extent of these changes. Summarising a review of definitions of transformation, Feola (2015, p. 377) suggests the term is broadly understood as a "major, fundamental change" whilst transition is something more "incremental" (see also Hölscher et al., 2018). There are also several discourses that can drive actor motivations on net-zero (Dryzek, 2005; Muir et al., 2000), as outlined in Table 2.1.

Some of the approaches echo the transformative aspects of planning theory, which seeks to use the planning process to transcend the status quo and contribute to post-neoliberal placemaking (Miraftab, 2009; Beard, 2003). Several authors have suggested that environmental discourse tends to be centred around (weak) ecological modernisation in the field of climate change (Hajer, 1997). Cowell suggests this shift is "massaging consent for, and thereby depoliticizing, fundamental questions about justice, markets, environments and the public sphere" (Cowell, 2013, p. 40). Meanwhile, grassroots actors have typically been excluded from higher climate governance—a challenge to reform neoliberal practices (Brown & Purcell, 2005). Indeed, social movement scholar Doherty suggests that a unified 'environmental movement' does not exist and there is a broad range of ideas, values and coalitions across the environmental sector, as reflected in Table 2.1 (see Doherty, 2007).

Instead, climate activism can be understood as embodying a movement containing diverse "convergent" spaces or networks which consist of networks at different scales that act independently, coalesce and then disperse (Routledge, 1996; Featherstone, 2008; Tompt, 2016). Understanding how transition dynamics play out at a local level and their linkages to wider phenomena can help explain planning's role in shaping transition pathways. For instance, in Scotland the devolved government has developed its own approach to inclusive, just planning approaches through its National Just Transition Planning Framework which it released in 2020 as part of the government's response to the Just Transition Commission's report (see Scottish Government, 2021).

As such, this chapter summarises the overarching UK environmental policy framework and the role of local government in achieving sustainable development and whether it indicates a shift from business-as-usual towards new ways of organising to encourage sustainable and inclusive agendas to emerge from the bottom up. This isn't an in-depth assessment of devolved government approaches to net-zero and nature-positive

Table 2.1 Shades of green in environmental discourses and related approaches to planning

Environmental Discourse	*Characteristics*	*Shade of Green*	*(Low-Carbon) Planning Examples*	*Authors*
Denial/ neoliberalism	Rejects climate change and environmental protection	None	Voluntary policies within existing market economy rules.	De Pryck and Gemenne (2017)
Green capitalism; ecological modernisation	Technology-focused, private sector orientated, voluntary arrangements within existing market economy	Light	Market shaping approaches. Expert-led. Planning to encourage service providers to provide low-carbon infrastructure; design codes, Community Infrastructure Levy (CIL - due to be replaced by the 'Infrastructure Levy' as per the 2023 Levelling Up and Regeneration Act).	Williams (2013)
Sustainable development	Non-Governmental Organisations (NGOs) and some government agencies	Shallow	NSDGs. Often NGO-led. Promotes circular resource usage; planner-led with experts and local consultation.	Berke and Conroy (2000) and Krizek and Power (1996)
Transition (sustainability transitions)	Gradual, imperceptible changes in societal structures	(i) Socio-technical, (ii) socio-institutional, (iii) socio-ecological	Gradual systemic shifts, e.g., in building standards.	Coenen et al. (2012) and Loorbach et al. (2017)

(*continued*)

Table 2.1 (continued)

Environmental Discourse	*Characteristics*	*Shade of Green*	*(Low-Carbon) Planning Examples*	*Authors*
Transformation	A major and often sudden fundamental change in societal structures; not always positive	Strong or disruptive	Soviet state breakdown. Brexit.	Feola (2015)
Deep ecology	Adherence to laws of nature, spiritual dimension (Gaia)	Deep—eco-fascist or eco-socialist	Community-led; self-reliance *Transition* movement. Rights of Nature movement (endowing legal personhood on nature).	Luke (2002)
Degrowth	Downscaling economic growth to levels compatible with environmental limits and the promotion of social well-being. Solidarity	Deep—reciprocity between natural and social worlds	Ecological Land Trusts. The Commons. One Planet Living (OPL). Doughnut Economics. Relocalisation 'parallel infrastructure'.	Escobar (2015) and Raworth (2018).
Just transitions	Fairness, redistribution of resources to promote social and economic equality and inclusion	Social inclusive redistribution	Allocation of resources and infrastructure to disadvantaged/ deprived communities; social housing.	Scottish National Just Transition Planning Framework (Scottish Government, 2021); Hughes and Hoffmann (2020)

Source: Adapted from Williams (2013) and Muir (2000)

approaches; however, we focus more on the English planning policy context under localism, and particularly the National Planning Policy Framework (NPPF) and how local governments can innovate for relocalised, or circular or even regenerative, economies within their planning and political architecture. We do this by first outlining the key roles and responsibilities of local government and their capacity to generate innovation in this area through formal and informal powers. We then go on to explore the role of the planning process as a key arena in which to critique how community-led actions intersect with existing policies and practices through a negotiated, yet politicised experience.

Democracy in Crisis or Opportunities to Consolidate Hybrid, Transition Management?

In some cases, social and environmental complexities are further compounded because globally trust in democracies is weakening. Research by scholars at the University of Cambridge indicates that global dissatisfaction with democracy has risen over time and has now reached the highest level, with the greatest frustration being in developed countries (Foa et al., 2020). These authors suggest that since 2005, democratic dissatisfaction has been on the rise in most countries and regions (except a small cluster of Scandinavian and some Asian countries), in what they term a "global democratic recession" (ibid., p. 2). As the chart below illustrates, despite a substantial fall in closed autocracies in the last two hundred years, these are now on the rise and other forms of democracy—electoral and liberal—are on the decline, and with those electoral autocracies (Fig. 2.1).

Similarly, according to the 2021 Edelman Trust Barometer[1] (central) government is perceived to be the least ethical and competent stakeholder (see also Davies et al., 2021). There were some small gains in the trust of national government as COVID-19 vaccinations programme were rolled out and the economy opened with tentative signs of success. Those dissatisfied with democracy may be tempted to support populist parties that seek to undermine liberal democracy (Pew Research Centre, 2018). This emphasises the critical need to get our relationship with democracy back on track, or to find solutions that address people's underlying dissatisfaction with the political system and what it tells us about our society, and even ourselves.

[1] https://www.edelman.com/trust/2021-trust-barometer

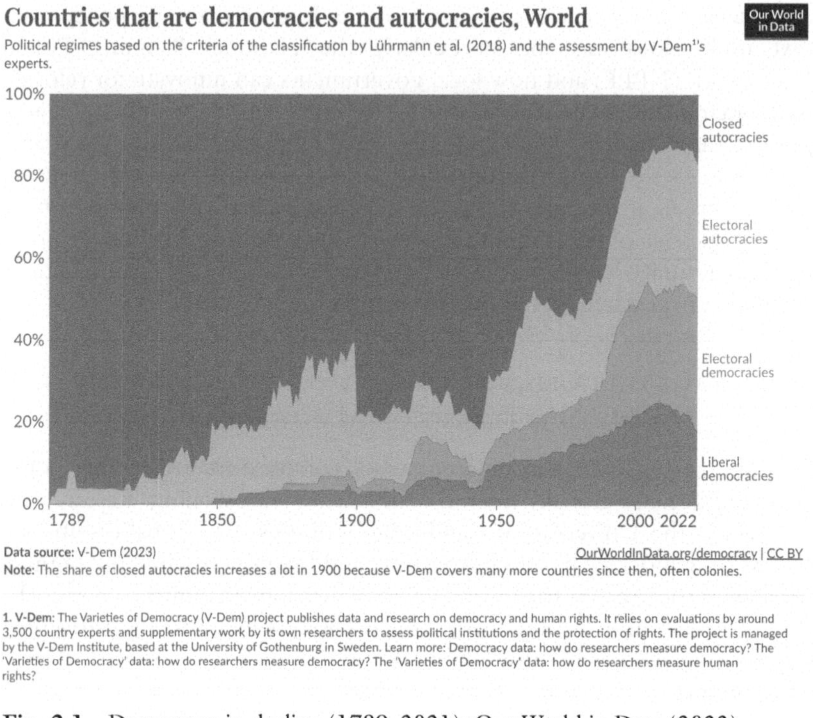

Fig. 2.1 Democracy in decline (1789–2021): Our World in Data (2022)

Recently there has been a plethora of local-level climate policies due to the more urgent pressures that have been caused through the decision to declare climate emergencies by several local councils. Yet a lack of clarity over how decentralised structures that link sub-national to national government relate to each other and the wider governing architecture compounds confusion over roles and responsibilities to deliver more sustainable solutions to key societal concerns (Shannon & Burrowes, 2021). In England, for instance, there is an increasing move towards unitary authorities that is changing the relationships between government, their supply chains and its citizens (Edelman, 2021). Indeed, after the immediate response to the COVID-19 pandemic, and with it the opportunities to reconfigure public services, "[m]ost governments did not pivot from traditional operating models to employ the agile, whole-of-government approach required for today's interconnected, rapidly evolving agenda" (Shannon & Burrowes, 2021).

Research suggests that it is at the local level where trust in local government has increased, and with it the belief in communities to deliver positive change (Thornborough, 2021). Localised green recovery programmes can also complement national efforts to act as an economic multiplier where investment in green infrastructure yields investment in jobs and new supply chains (Shannon & Burrowes, 2021). Yet, while local governments may aspire to adopt innovative or alternative approaches, there is often a shortage of particular expertise and staff time to do so. As a recent survey by the National Association of Local Councils (NALC, 2021) found, this is particularly the case in relation to climate action and having the confidence to feel that local councils at the town and parish level can make a difference, owing to the fact they have limited formal powers.

Net-Zero and Climate and Ecological Emergencies—the UK Government's Policy Response

A government's attitude towards sustainable development is reflected in what objects are obligatory and which are discretionary (Cowell, 2013). The UK is also required by law to comply with international agreements to reduce carbon emissions through associated carbon budgets that incrementally reduce the number of emissions generated. The amended 2008 Climate Change Act committed the UK government to reduce greenhouse gas emissions to net-zero by 2050 (compared to 1990 emissions levels) as part of the Paris Agreement. Originally, this was an 80% reduction of emissions in this period. The UK government has committed to a net-zero reduction target of at least 68% by 2030 through its Nationally Determined Contributions (NDCs)—the highest of any other global economy, which was staged to increase to 78% reduction in emissions by 2035, in line with the government's sixth carbon budget. The Conference of the Parties (CoP) in Glasgow in November 2021 also sought ambitious greenhouse reduction targets for 2030, which many climate and ecological emergency commitments are working towards.

Some of the low-hanging decarbonisation fruit has been picked, with the UK meeting earlier carbon budgets; that is, the first (covering the period 2008–2012) and the second (covering the period 2013–2017) have been met. The UK is on track to meet the third carbon budget but is not on track to meet the fourth or the fifth (Table 2.2). A Climate Change Committee (CCC) report stated that the government is making partial progress in 15 out of 31 milestones (namely transport and industry). For

Table 2.2 UK carbon budget obligations and progress

Carbon Budget	*Period*	*Met/On Track?*
First	2008–2012	Met
Second	2013–2017	Met
Third	2018–2022	Met
Forth	2023–2027	Uncertain
Fifth	2028–2032	Uncertain

others, this is a policy work in progress (CCC, 2020). As of 2019, little progress was made on 21 indicators, and in some cases, these worsened. Recent 2023 updates from the CCC continue to stress that insufficient action is being taken at the levels required to achieve the government's net-zero targets.

Increasingly, corporate companies are required to report on their net-zero impacts, a condition that is starting to trickle down their supply chains and affecting smaller-scale small and medium-sized enterprises (SMEs). There remains much more work to be done on the nature-positive elements of sustainable business though there are some encouraging advances in this, including a Defra 'Nature Positive' consortium with business and other stakeholders and the Green Finance Institute's 'Hive' which showcases resources and best practice on green finance.[2]

Under the UK's forthcoming Sustainability Disclosure Standards, companies with over 500 employees, asset managers and financial institutions will be required to report their Environmental, Social and Governance (ESG) impacts and performance. The Standards will be phased in, with some elements coming into force in 2024 and the full framework expected to be operational by 2025. The SDR is also expected to be closely aligned with the Taskforce for Nature-Related Financial Disclosures (TNFD), the Taskforce for Climate-Related Financial Disclosures (TCFD) and the International Sustainability Standards Board (ISSB) standards (investor-focused measures that aim to integrate various standards and align these to accountancy practices). (In comparison, the EU's reporting requirements are required for companies over 250 employees in the EU). The UK's Green Taxonomy (mirroring its EU counterpart) aims to accelerate green financial investment by detailing the granular definition of green economic activities that companies can report against. This will bring with it unprecedented regulation relating to mandatory carbon reporting and net-zero

[2] https://www.greenfinanceinstitute.co.uk/gfihive/

accounting for products and services in some sectors, notably the financial services. However, while the EU and even Indonesia have announced their taxonomies to determine as yet the UK is yet to formally announce its Green Taxonomy (though it is in its final stages of development).

Planning and construction is a key sector that requires improved greening outcomes: due to government policy, houses are being built today that will need to be retrofitted in the future to make them compatible with the UK's commitment to reach net-zero by 2050. The UK has one of the worst energy efficiency standards in its building stock in Europe, accounting for as high as 40% of UK emissions, with transport as the largest emitting sector with energy following behind. COVID-19-related emissions from reduced activity and travel during the pandemic led to a fall of 6.4% of emissions globally in 2020 (Tollefson, 2021). However, these cuts were only temporary, and the war in Ukraine poses yet another seismic challenge and opportunity to rethink and rescale our energy systems. The cost-of-living crisis, however, may reduce the impetus for households to support net-zero investment when it adds to their already inflated energy bills. To achieve net-zero, the UK needs to achieve an average emissions reduction of around 15.5 $MtCO_2e$ (million tonnes of carbon dioxide equivalents) per year over the next 30 years (CCC, 2022). Now that the UK has committed to net-zero by 2050, it is even more critical that progress on meeting the fourth and fifth carbon budgets is accelerated.

On the 'ecological' aspects of emergencies, the 2021 Environment Act was heralded as a "step change" by the UK government due to its introduction of pioneering forms of environmental management, such as biodiversity net gain (BNG—leaving the environment and habitats in a better condition than before development), which was expected to come into force in November 2023 and (as of September 2023) was pushed back (again) into 2024 (or 2025 for national infrastructure projects). In anticipation of BNG, many government institutions, local government and environmental records centres have begun to reorient governance to deliver on net gain, even before the Environment Act was passed. Though, as yet, no mandatory reporting commitment exists for nature and biodiversity in the same way it does for larger businesses reporting on carbon impacts, which further entrenches the divide between climate and ecological emergency responses (see Owen et al., 2022). Though the release of the Outcomes Indicator Framework in January 2023 may help to partially bridge a data gap on environmental impacts some environmental NGOs suggest that government is falling woefully behind delivering its policy ambitions for nature (Mittal, 2023).

The UK Government department Defra has been active in promoting environmentally focused stewardship schemes for farmers and landowners in the context of post-Brexit environmental policy as the government withdraws from the EU's Common Agricultural Policy (oft-criticised for paying farmers simply for owning land rather than enhancing nature on their land). The Environmental Land Management Scheme (ELMS) encourages improved environment and biodiversity through a grant payment scheme that rewards farmers and landowners for positive environmental measures. Meanwhile, the Agriculture Act 2020 reframes environmental discourses underpinning agricultural land management, emphasising the delivery of 'public good' benefits (such as improved soil and water quality). Although the Agriculture Act 2020 doesn't extend into the realm of urban and spatial planning, the concept of public goods could help reframe the purpose of planning.

However, while this is a significant step forward there are still key issues to iron out (see also Mittal, 2023), such as how different types of landowners and farmers (such as tenant and small-scale family farms) will be equally rewarded and how private markets that fund offsetting schemes will complement publicly funded schemes, such as ELMS (see Lyon & Burnett, 2023). A more fundamental issue is the disconnect between agricultural policy and planning policy—formalised through the 1948 Town and Country Planning Act. A lack of a joined-up approach hampers more integrated land-use solutions, which has led the Food, Farming and Countryside Commission (FFCC) to call for a Land-Use Framework to solicit a holistic response to the most appropriate land use across a range of land-use types (FFCC, 2022), which despite being promised by government still remains unpublished (Mittal, 2023).

Localised Responses to Climate and Ecological Emergencies

Globally, climate emergency declarations have been declared in 2355 jurisdictions spread across 40 countries that covers 1 billion (bn) citizens (as of February 2024). According to NewClimate Institute, this covers 826 cities, 103 regions and 1565 companies in every continent.[3] Over 95% of the UK population is covered by a council that has declared a climate emergency.[4]

[3] See: https://www.theclimatemobilization.org/world-map.

[4] https://climateemergencydeclaration.org/climate-emergency-declarations-cover-15-million-citizens/

Data compiled by Climate Emergency UK in February 2021 states that 300 out of 408 (74%) District, County, Unitary and Metropolitan Councils have declared a Climate Emergency, as well as eight Combined Authorities or City Regions.[5] As of February 2024 this had risen to 570 councils.[6]

However, a recurrent lack of global, or even national, action on climate change has led many at the 'local' level to seek radical, grassroots action on climate change—often resulting in pioneering bottom-up policies. For instance, in 2017—well before the mainstream call to declare a climate or ecological emergencies—the market town of Frome, the principal case study of this book, was one of the first to declare a commitment to go "fossil free" by 2040.[7] This was mirrored at a national level in June 2019 through an amendment to the 2008 Climate Change Act, where national government set a target of net-zero emissions by 2050, compared to the previous target of at least an 80% reduction by that date.

The UK's first citizens' assembly on climate change, Climate Assembly UK,[8] held its first meeting in January 2020 in the city of Birmingham; this was commissioned by six cross-party House of Commons Select Committees in summer 2019, in response to the UK Government's commitment to meet net-zero carbon emissions by 2050. The assembly was made up of a group of Expert Leads, an Advisory Panel and an Academic Panel; these bought together climate specialists, business leaders, constitutional and economic experts, civil society organisations and 110 members of the public, aiming to be a representative panel to enable the public to have their say about how the UK could shift to a net-zero economy.

Local authorities and local councils are seeking to create comprehensive, action-oriented and deliverable Climate Action Plans, giving rise to forms of new environmental governance. This is critical if local action is to be coupled with coordinated action at a national and global level to ensure that target setting is grounded in a plan for its delivery at different scales. Many climate action plans integrate principles of inclusive planning, reflecting the notion that climate action should be based on fair and

[5] https://www.climateemergency.uk/blog/list-of-councils/

Arup has prepared guidance for councils that have declared a climate emergency (Arup, 2019).

[6] https://climateemergencydeclaration.org/climate-emergency-declarations-cover-15-million-citizens/ See: https://www.theclimatemobilization.org/world-map

[7] https://ecohustler.com/culture/frome-council-plans-to-be-the-uks-first-fossil-fuel-free-town

[8] https://www.climateassembly.uk/

equitable policies, such as those reflected within 'just transition' discourses (Ciplet & Harrison, 2020; Martin et al., 2020).

This is mirrored through participatory approaches that are being invoked to elicit input from different stakeholders, such as climate citizen assembles and localised fora, such as the Net Zero Assembly in Devon, which has created the Devon Carbon Plan using citizen assemblies to elicit input from residents, community groups and businesses.[9] Local governments are key to ensuring an accountable and actionable approach for ideas suggested by local citizen assemblies. The Devon Climate Emergency Partnership includes all local government in the county and has committed to responding to the recommendations arising from the local citizens assembly and will carry forward the implementation of the Plan. In Birmingham, a 'Route to Zero (R20)' Taskforce was created in Autumn 2019 and brings together the City Council and representatives from the West Midlands Combined Authority, the National Health Service (NHS), higher education, the business community, faith communities, young climate strikers, climate campaigners, and other key partners and stakeholders.[10]

In Bristol, a One City Approach[11] has been developed as a partnership-based model of governance to encourage a fairer, healthier and sustainable city. This has involved creating a One City Plan[12] setting out challenges and solutions in reaching city-wide goals by 2050. These plans are supported by the Bristol Green Capital Partnership which oversees the One City Environment Board and the Bristol Advisory Committee on Climate Change. To align local action with the Sustainable Development Goals (SDGs), strategic partnerships have been created to forge the Bristol SDG Alliance, including introducing measures to monitor progress towards them and partnerships ensuring accountability to their delivery. In 2016, the London Borough of Sutton committed to becoming a One Planet Living (OPL) borough by 2016. The London borough of Harrow has developed a pan-London 'One World Living' Action Plan to reduce unsustainable consumption behaviours as part of the London Council's wider climate change programme (although this isn't explicitly linked to OPL principles).

[9] https://www.devonclimateemergency.org.uk/devon-carbon-plan/citizens-assembly/

[10] See https://www.birmingham.gov.uk/info/20015/environment/2026/climate_emergency/4

[11] https://www.bristolonecity.com/

[12] https://www.bristolonecity.com/about-the-one-city-plan/

Some progressive councils are going beyond the SDGs and seeking innovative sustainability frameworks to reorientate decision-making to integrate sustainability and well-being within the context of planetary boundaries. For instance, Cornwall Council is using doughnut economies (Raworth, 2018) to account for multiple impacts of climate in/action on council decision-making.[13] The doughnut model is also being applied in rural contexts to promote sustainable tourism, such as the Park Doughnut in the Brecon Beacons National Park Authority.[14] In Wales, since 2010 the government has supported the 'One Planet Development' framework which calculates the environmental impact of development by measuring its ecological footprint. This approach included setting up a One Planet Development Council which acts as an intermediary between applicants and local planning authorities and provides additional guidance on development issues. A 2022 report by the One Planet Council found the One Planet Development model was, encouragingly, exceeding expectations (One Planet Council, 2022).

As we will explore more in this book, environmental social movements initiatives and other environmental actors in England are increasingly seeking to becoming their own developers using the rights under localism through the creation of bottom-up Neighbourhood Plans or using Community land trusts (CLTs) to support community-led, affordable and sustainable development.

These are many clearly innovative, locally rooted responses available to encourage an inclusive, planned approach to climate and ecological action. However, not all climate emergency declarations have been centred on a robust participatory approach. For instance, Howarth et al. (2021) suggest that there are four main 'journeys' to declaring a climate emergency: (i) active from above, (ii) passively from above, (iii) active from below and (iv) passively from across. That is, the intention can be either a political gesture, genuine attempts to stimulate bottom-up activity or mere statements of intent, devoid of any meaningful strategy.

The Structure of Local Government

In England, as of April 2023 there are a total of 317 local authorities, which are made up of either 'single-tier' (unitary) councils or two-tier councils (smaller districts and one larger county council, which have split

[13] See https://vimeo.com/476197985

[14] https://www.beacons-npa.gov.uk/the-authority/evidence-and-research/

local government functions between them). There are currently 21 county councils which typically cover historic and rural counties and are responsible for strategic matters (but not planning) and 164 district councils. Typically, there are between five and seven district councils for each county council, with the former and relating to place-based services, like planning and housing. Single-tier councils include the 36 metropolitan boroughs (e.g., outer city London areas, larger cities, some suburbs and rural areas), 62 unitary councils and the 32 London boroughs (see DLUHC, 2023). There are also 11 combined local authority areas and over 10,000 parish and town councils.

The Local Government Association (LGA) represents local government from the district level and above. The National Association of Local Councils (NALC) represents town and parish councils in England as well as community councils in Wales. (For more information on the administrative make up of English councils there is a great summary on the Local Government Information Unit website or there's more on the Gov.uk site 'Understand how your council works', or see DLUHC, 2023).

Approximately 30% of England is 'parished' that is, falls within a town or a parish council (NALC, 2021), this means that two-thirds of England is not represented by a town or parish council. In these cases, the district or unitary council is the closest residents and citizens get to a local representative government body. NALC have suggested that this is causing massive divides in democratic representation, where there can be cases of 12,500 residents for every councillor (as in the London Borough of Newham) compared to rural parished areas, such as 25 residents per parish councillor in Weeford, Staffordshire.[15]

Northern post-industrial 'red wall' communities are also less likely to be parished, causing democratic implications for the government's intention to 'level up' economic and social regeneration in these areas. The political gains made under Boris Johnston in the 2019 general election in communities in the Midlands and the North of England is waning. The May 2024 local elections saw a return to the Labour vote in key red wall areas such as Blackpool South and the election of a Labour Mayor for the East Midlands. However, many of these seats were highly contested and also saw a steady rise by the Green Party as well as the independent vote—which suggests a fracturing of political identities in UK politics.

[15] https://www.ukonward.com/reports/double-devo-town-and-parish-councils/

While 270 new local councils have been created since the passing of the Localism Act in 2011, these have often been the result of the formation of unitary governments (which would by implication absorb district councils) or local government restructuring, not necessarily from a bottom-up desire for communities to seek to establish more representative, local entities. In some cases, such as Tower Hamlets in 2019, proposals by local communities to become parished have been rejected by local authorities.

Local Government and the Hangover of Austerity Localism and COVID-19 Recovery

Another key issue facing local governments is the hangover from what Lowndes and Pratchett (2011) term 'austerity localism', that is the austere resourcing constraints imposed on them by central government by the 2010–2015 Conservative Lib–Dem Coalition and beyond. The LGA (LGA, 2018) estimated that by 2020, austerity measures led to a reduction in core funding from central government to local authorities by nearly £16bn, compared to the previous decade. This equates to a loss of 75p out of every £1 by 2020 (LGA, 2017). In 2014, the National Audit Office projected a 37% decrease in funding from 2010–2011 to 2015–2016 in local authority funding (NAO, 2014). In 2020, 168 councils received no revenue or grant support from national government and services provided by local government will face a shortfall of £7.8bn by 2025 (LGA, 2018). This has been exacerbated by the effect of the COVID-19 pandemic, where UK councils are now estimated to be in an even more precarious position of having a £4bn 'black hole' which is affecting many councils' ability to carry out statutory duties. One of these functions is to deliver a workable and effective planning system: within such a context the value of planning is weakened when its regulatory role is increasing hollowed out by the state.

Some councils are on the verge of bankruptcy and are increasingly looking to central government to fill a shortfall in income. Comparatively, local governments in the South West are expected to make a lesser degree of cuts than London and other regions (BBC, 2021). The "Barnet graph of doom"—when spending overtakes income received, meeting on a downward income forecast trajectory—depicted that the London borough's entire projected budget would be consumed by children's' services and adult social care (see also OBR, 2014). As a response, some councils moved towards outsourcing of key statutory functions (e.g., Barnet) to private third parties, compromising local governments traditional function to deliver local social and public services. Though, Barnet Council—which had outsourced a significant proportion of its service delivery—is now seeking to bring key services (including planning and highways) back in-house.

Unfortunately, despite the progress in championing environmental policy, in 2023 Birmingham City Council—which serves 1.14 million people—declared itself effectively bankrupt following a costly fair pay case at an employment tribunal and added costs from an IT system rollout. The Council now faces a £300 million black hole, which it hopes to partially plug with a 10% hike in council tax in 2024–5. Other local authorities have also signalled they are not far from financial collapse (which we further discuss in Chap. 11), including some councils from the most deprived areas of the UK. Amid the issue of protracted austerity measures, Brexit and the cost-of-living crisis, delivering on an innovative agenda programme that can address climate change and ecological breakdown is one of the many competing and urgent pressures facing local government. Whether independent politics can support to help underpin the foundations for more radical, community-led action is a key inquiry of this book.

Party Politics and the Clamour for Political Alternatives

Extending a trend for greater levels of self-government, arguably dating back to the reformist agenda (Wills, 2016), there has been an increasing appetite for distributed powers at the local and regional level, indicated by the Scottish Referendum and notably Brexit which, while seeking national 'freedoms', is indicative of the wider sense to 'take back control' from seemingly remote political institutions. These trends may pave the way for a surge in new conversations about (net-zero) political post-Brexit alternatives.

Yet despite moves under localism to devolve power, the UK is still considered as one of the most top-down and restrictive national state governance arrangements in the West (Hodson & Marvin, 2013; Ehnert et al., 2018). The 2022 Levelling-up White Paper which sought to 'empower local leaders and communities' mentions parish councils only twice and town councils not at all (which might be because town councils are technically parish councils that have decided to call themselves town councils).

Furthermore, Cowell argues that with the abolition of regional government the political spaces to ask what the most sustainable solution for a region or nation is curtailed; "[t]he best answers may not equate to the aggregate consequences of local determinations", and varied interpretations pose a challenge to coordination (Cowell, 2013, p. 34).

Onward has proposed a 'double devolution', whereby communities without a local council can decide by automatic ballot if they want to establish one and "explore a radical expansion of neighbourhood control through local (parish and town) councils" (Treadwell et al., 2023). This would involve ensuring that at least two-thirds of local councils are elected (in some cases, due to a lack of standing candidates individuals are co-opted into the council, without being elected). NALC also suggest that neighbourhood councils should be able to 'pull down' responsibilities from local authorities if communities perceive them to be under-delivering or if local actors feel they could do these duties in a better way. This would massively extend the responsibility, and autonomy, of communities. These measures are reflective of ***municipalism***, where a city or town or municipality has greater powers to govern.

Meanwhile, the UK political system is adversarial whereby party politics actively pitches those from different political persuasions against one another (the House of Commons chamber is an obvious case in point). Smaller parties are under-represented because of the first-past-the-post system which rewards candidates that win seats based on a simple majority (even in constituencies with far fewer seats than some urban areas). A previous attempt to force change with a referendum on a proposed Alternative Vote, a voting system that would give minority parties a better entry point, was voted down by 67.9% of the UK electorate in 2011.[16] This arguably had the effect of shrinking political debate "when there is a growing clamour to expand it", particularly towards debates on post-Brexit institutional reform (Nandy et al., 2016, p. xiii). Efforts by some Members of Parliament (MPs—representatives of the lower chamber of the UK national legislature, the House of Commons) and the political action charity Compass are actively seeking 'progressive alliances' between MPs to encourage socially responsible and environmentally sensitive UK politics, encouraging them to put aside party politics for the greater national good. New forms of environmental governance may offer new routes to vocalising alternatives democratic ambitions.

[16] https://www.electoral-reform.org.uk/voting-systems/types-of-voting-system/alternative-vote/ (last accessed 19.1.19).

Planning for Climate Emergencies—Action and Ambition at a Local Level

The 2019 amendment to the Climate Change Act (2008) amendment in 2019 sets out a net-zero target but there have been no clear and mandatory frameworks recently for local governments to report against and even less clarity or resource at the level of local councils, that is, town and parish councils (the lowest tier of political administration in England). Though this could be set to change with the January 2023 Environmental Improvement Plan and the 66 associated outcome indicators to measure performance against the 25-Year Environment Plan, towards much needed coherence in the Outcome Indicator Framework.

As we shall explore below, the UK government has been active in dismantling the functions of the state by removing various powers from local government to deliver on low-carbon, or net-zero, outcomes. Yet Hodson and Marvin have suggested that deregulation represents the active reshaping of sub-national governance (under the umbrella of localism) which creates competition for limited resources. Low-carbon action is to 'emerge' through institutional mediators that can support entrepreneurialism rather than through specific policy measures (Hodson & Marvin, 2013, p. 82). Delivering climate change ambitions amid constrained resources, in theory, invites communities to play a greater role in their delivery on local-level actions to find solutions with local government to deliver on climate pledges.

Both NALC and the LGA have committed to supporting climate emergency declarations. NALC ran a survey (which had 211 responses) between October 2020 and January 2021 to find out to what extent local councils were addressing climate change, and what barriers prevented them from doing so (NALC, 2021). It found those making declarations were more likely to follow through on sustainability measures, such as sustainable neighbourhood plans. However, much more is needed to be done to help local councils baseline their emissions and environmental assets. Those with higher budgets tended to be more likely to make declarations and councils indicated they wanted more resource and support in this area.

The LGA has set up a climate change hub to support councils represented by them.[17] In January 2021, NALC also set up a climate emergency group to promote action on the ground, including a series of meetings for parish and town councillors and a WhatsApp group to enable communication between councillors and other stakeholders that joined the

[17] https://www.local.gov.uk/our-support/climate-change-hub

group. Other important materials to support local government to 'transition' towards more sustainable forms of governance and service delivery have been developed, such as the LGA's Councillor's workbook on the *Local Pathway to Net-Zero* (LGA, 2021).

In addition, I carried out my own survey between November 21 and February 2022 which sought to understand the extent to which climate action was being measured, what barriers there were but also what design features were needed to help support different actors between monitoring and evaluating climate action. It also asked respondents' views on the extent local and neighbourhood plans were able to deliver on climate action.[18] It had a relatively low number of respondents as surveys go, mainly parish and town councils, but the results are indicative of how net-zero commitments are being addressed.

Overall, 35% (20) respondents said that sustainability covered a "moderate amount" of their role, with 28% of respondents (16) said that it was central to everything they did in their job, covering more explicit environmental roles described above. A total of 19% (11) respondents said that it covered most of what they did, but, equally, 19% (11) respondents said that sustainability only covered a little or indeed none of the activities in their roles. At that time, 37% of respondents had made a climate emergency declaration and only 8% had made a biodiversity (ecological) declaration. Thirty-five percent of respondents had not made or were not planning on declaring either a climate or an ecological emergency. Typically, it was the more resourced local councils making these delcarations.

The issues that most affect respondents' ability to monitor and evaluate sustainability within their role or organisation were:

- Lack of funding
- Lack of appropriate tools and the time to learn new ones
- Not being aware of appropriate tools
- Not having a one-stop shop for tools, and so on
- Unsupportive policy context
- Lack of alignment to SDGs
- Lack of training
- Too many conflicting priorities
- Not being seen as a priority at an organisational level

[18] You can view the results and analysis to my survey here: https://express.adobe.com/page/TyM7eOfNYmmxH/

Thinking Creatively About Democratic Action and Incentives at the Local Scale

There are calls for legislative reform to further enhance the potential for participation in UK climate policy. The Climate and Ecological Emergency (CEE) Bill (currently in its second reading in House of Commons) seeks to require the UK government to meet its climate and nature targets.[19] The Bill was proposed by the Green Party MP and former leader Caroline Lucas, and heavily campaigned for by Extinction Rebellion and has cross-party support across the major parties. It proposes the introduction of a Climate and Nature Assembly (like Citizens Assemblies) to advise the Secretary of State to develop a comprehensive strategy to meet such targets and to strengthen the CCC to deliver on these ambitions. The Office of Environmental Protection (OEP), established under the Environment Act 2021, performs a similar function[20] in England and Northern Ireland and an MoU in place for how it works with environmental oversight bodies in Wales and Scotland (see Macroy, 2023), but it has been criticised for not being independent from the government and does not have a central bottom-up thread to its governance structure. The OEP has played a role in challenging the government over its failure to deliver on its pledges to tackle nature-based degradation and poor air quality. While providing an important oversight function, the OEP has been criticised for not being independent from the government, and nor of having a 'bottom-up' thread to its governance structure. There are concerns that the OEP may struggle in its duty to keep government accountable on its compliance with environmental law—itself pointing out that government is failing to always provide legally-required Post-Implementation Reviews (PIRs) of environmental laws (OEP, 2023).

In October 2021 the parliamentary select committee for Levelling Up, Housing and Communities published a report - with a name very close to the LGA's from the same year—the *Local Path to Net-Zero* which suggested that government needs to work with local government to deliver a net-zero delivery framework (HoC Select Committee, 2021). This report sets out the roles and responsibilities of local and central government and clarifies the potential role of local councils in delivering a just transition framework which will require clear, long-term funding routes—funding

[19] https://bills.parliament.uk/bills/2943

[20] The OEP covers England and Northern Ireland but it has a remit on 'reserved matters' which are the responsibility of UK government rather than devolved administrations.

that has yet not been made explicit in recent government policy. To deliver on net-zero ambitions (summarised below), government needs to work with local councils and communities to devise and implement policies and actions relating to housing and planning, energy and the promotion of equitable and inclusive outcomes (see *just transitions*—see Table 2.1). These opportunities are also echoed by the LGA in its various resources to local government on net-zero transitions.[21]

More recently, the 2023 Levelling Up and Regeneration Act seeks to address geographic disparities within parts of the UK and to spread opportunities 'equally', with associated targets for 'levelling up'. To do so it aims to reform the planning process and encourage further devolution of powers at the local level, enable more affordable housing and (regionally-strategic) loans to SMEs to support regeneration in priority areas. However, while it prioritises green regeneration, particularly on brownfield sites, the Bill was criticised for not sufficiently detailing its contribution to sustainable development (Newberry & Dingemans, 2022).

In addition, as I explain elsewhere, there is a need for mechanisms that encourage incentives that all work in tandem to encourage sustainable outcomes, or what I refer to as *regenerative value* (Burnett, 2022a; see also Toledo, 2023). These mechanisms could link to citizens' assemblies, monitoring and indexing of actions (see Conclusion—Towards regenerative development in an era of uncertainty? and Burnett, 2022b for more on this as it relates to the planning process). The publication of the UK government's Environmental Outcome Indicators Framework - which sets out indicators (66 as of June 2023) to measure the ten objectives of the 25-Year Environment Plan—together with environmental movements that seek more inclusive engagement with local democratic parallel infrastructure (more of that in the final chapter), could indeed lay the foundations to operationalise a sustainability-focused and socially inclusive regenerative value.

In the survey I carried out in 2021, out of possible options presented the most popular potential features to help improve how respondents approach sustainability in their roles included (see Bream-McIntosh et al., 2023 for a discussion of how multi-level sustainability platforms can further enable some of these ideas)[22]:

[21] https://www.local.gov.uk/net-zero

[22] You may also wish to refer to a presentation I gave to NALC and local councillors about potential solutions: https://express.adobe.com/page/BR88Ce9DbhmDs/

- Peer-to-peer exchange on sustainability issues.
- An Ideas Bank to log new ideas and help resource them.
- Benchmarking my/my organisation's progress on sustainability.
- A reporting function, for example, for customised quarterly or annual reporting.
- A searchable database of sustainability tools and approaches.
- A single digital platform that consolidates different sustainability reporting tools.
- An ability to log my actions according to customised organisational or personal indicators.
- An ability to log my actions according to national or regional indicators.
- Values mapping function, that is, seeing what kinds of approaches, people and partners I might work well with.
- Indexing my/my organisation's progress against others in the same field.

A potential way that different elements of regenerative value could reinforce each other to promote sustainable outcomes is depicted in Fig. 2.2. We return to some of these ideas in the final chapter after assessing how

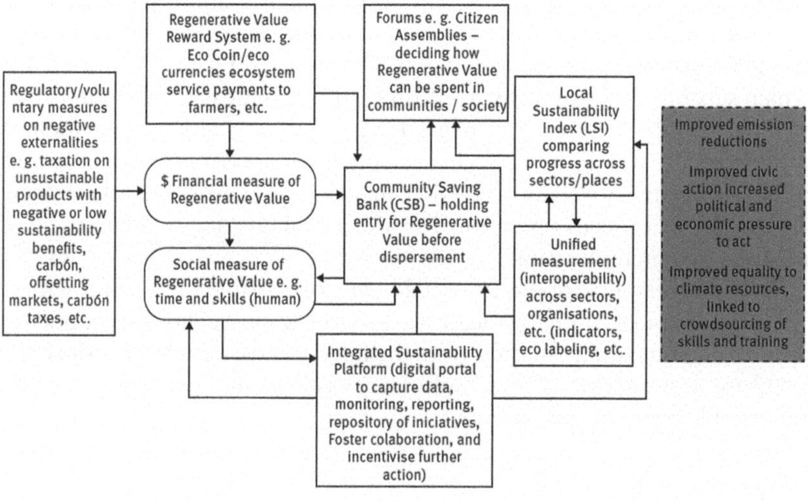

Fig. 2.2 The lifecycle of the creation and redistribution of regenerative value (Burnett, 2022a, p. 184)

conditions of placemaking, taking Frome as an example, might affect how these solutions might play out and evolve to support more inclusive, radical and effective place-based action in the future.

However, the key to responding to the call for more participatory governance is the capacity of grassroots actors to carry out such duties and the resources available to them. Parish and town councillors are voluntary roles (district and unitary councillors get some financial reimbursement for their duties while MPs get a paid salaried position). Who steps up to be a local councillor can set the tone for which outcomes follow. The politics of such transition—through planning and placemaking—and how the ideals of participatory climate action hold up in the context of navigating complex political and planning institutions is the focus of the rest of this book.

References

Allmendinger, P., & Haughton, G. (2012). Post-political spatial planning in England: A crisis of consensus? *Transactions, 37*(1), 89–103.

Arup. (2019). You've declared a climate emergency…what next? Guidance for local authorities. https://www.arup.com/perspectives/publications/promotional-materials/section/youve-declared-a-climate-emergency-what-next-guidance-for-local-authorities

BBC. (2021). Covid leaves UK councils with £3bn financial black hole. Retrieved July 9, 2021, from https://www.bbc.co.uk/news/uk-57720900

Beard, V. A. (2003). Learning radical planning: The power of collective action. *Planning Theory, 2*(1), 13–35.

Berke, P. R., & Conroy, M. M. (2000). Are we planning for sustainable development? An evaluation of 30 comprehensive plans. *Journal of the American Planning Association, 66*(1), 21–33.

Bream McIntosh, O., Burnett, A., Feldman, I., Lamphere, J., Reuter, T., & Vital, E. (2023). The role of sustainability knowledge-action platforms in advancing multi-stakeholder engagement on sustainability. *Data & Policy, 5*, E34. https://doi.org/10.1017/dap.2023.31

Brown, J. C., & Purcell, M. (2005). There's nothing inherent about scale: Political ecology, the local trap, and the politics of development in the Brazilian Amazon. *Geoforum, 36*(5), 607–624.

Burnett, A. (2022a). Operationalising performance measurement and incentives for 'regenerative value' to improve net sustainability outputs. In G. Vélez-Cuartas & O. Romero-Goyeneche (Eds.), *Transformative metrics: Contributions to the studies for monitoring and evaluating how science, technology, and innovation can address social and environmental challenges.* University of Antioquia.

Burnett, A. (2022b). Securing a regenerative and just planning system. *Town & Country Planning, 91*(2), 123–128.

CCC. (2020). 2020 progress report to Parliament. Climate Change Committee. https://www.theccc.org.uk/publication/reducing-uk-emissions-2020-progress-report-to-parliament/

CCC. (2022). Reaching net zero in the UK. Retrieved June 4, 2023, from https://www.theccc.org.uk/uk-action-on-climate-change/reaching-net-zero-in-the-uk/

Ciplet, D., & Harrison, J. L. (2020). Transition tensions: Mapping conflicts in movements for a just and sustainable transition. *Environmental Politics, 29*(3), 435–456.

Coenen, L., et al. (2012). Toward a spatial perspective on sustainability transitions. *Research Policy, 41*(6), 968–979.

Cowell, R. (2013). The greenest government ever? Planning and sustainability in England after the May 2010 elections. *Planning Practice & Research, 28*(1), 27–44.

Davies, B., et al. (2021). Changes in political trust in Britain during the COVID-19 pandemic in 2020: integrated public opinion evidence and implications. *Humanities and Social Sciences Communications, 8*(1), 166.

De Pryck, K., & Gemenne, F. (2017). The denier-in-chief: Climate change, science and the election of Donald J. Trump. *Law and Critique, 28*(2), 119–126.

DLUHC. (2023). Local government structure and elections. Department for Levelling Up, Housing and Communities. Retrieved from https://www.gov.uk/guidance/local-government-structure-and-elections#structure

Doherty, B. (2007). Environmental Movements. In C. Okereke (Ed.), *The Politics of the Environment*. London, UK: Routledge.

Dryzek, J. S. (2005). *The politics of the Earth*. Oxford University Press.

Edelman. (2021). *2021 Edelman Trust Barometer UK Findings. Edelman Trust Barometer*. Retrieved from https://www.edelman.co.uk/edelman-trust-barometer-2021-uk-findings

Ehnert, F., et al. (2018). Urban sustainability transitions in a context of multi-level governance: A comparison of four European states. *Environmental Innovation and Societal Transitions, 26*, 101–116.

Escobar, A. (2015). Degrowth, post-development, and transitions: A preliminary conversation. *Sustainability Science, 10*(3), 451–462.

Featherstone, D. (2008). *Resistance, space and political identities: The making of counter-global networks*. Wiley.

Feola, G. (2015). Societal transformation in response to global environmental change: A review of emerging concepts. *Ambio, 44*(5), 376–390.

FFCC. (2022). *A land use framework for England*. Food Farming and Countryside Commission. https://ffcc.co.uk/publications/a-land-use-framework-for-england

Foa, R. S., et al. (2020). *Global satisfaction with democracy 2020.* Bennett Institute for Public Policy.

Hajer, M. A. (1997). *The politics of environmental discourse: Ecological modernization and the policy process.* Oxford University Press.

Hodson, M., & Marvin, S. (2013). *Low carbon nation?* Routledge.

Hölscher, J., et al. (2018). Transition versus transformation: What's the difference? *Environmental Innovation and Societal Transitions, 27*, 1–3.

Howarth, C., et al. (2021). What next for local government climate emergency declarations? The gap between rhetoric and action. *Climatic Change, 167*(3), 27.

Hughes, S., & Hoffmann, M. (2020). Just urban transitions: Toward a research agenda. *Climate Change, 11*(3), e640.

Krizek, K., & Power, J. (1996). *A planner's guide to sustainable development.* American Planning Association.

Leeuwis, C. (2000). Reconceptualizing participation for sustainable rural development: Towards a negotiation approach. *Development and Change, 31*(5), 931–959.

LGA. (2017). *Local government finance and arrangements beyond 2020.* Local Government Association. https://www.local.gov.uk/parliament/briefings-and-responses/lga-briefing-local-government-finance-and-arrangements-beyond

LGA. (2018). *Local government funding—Moving the conversation on Finance.* Local Government Association. https://www.local.gov.uk/sites/default/files/documents/5.40_01_Finance%20publication_WEB_0.pdf

LGA. (2021). A Councillors Workbook on the Pathway to Net-Zero. *Local Government Association.* https://www.local.gov.uk/publications/councillor-workbook-local-path-net-zero

Loorbach, D., et al. (2017). Sustainability transitions research: Transforming science and practice for societal change. *Annual Review of Environment and Resources, 42*, 599–626.

Lowndes, V., & Pratchett, L. (2011). Local governance under the coalition government: Austerity, localism and the 'Big Society'. *Local Government Studies, 38*(1), 21–40.

Luke, T. W. (2002). Deep ecology: Living as if nature mattered. *Organization & Environment, 15*(2), 178–186.

Lyon, F., & Burnett, A. (2023). Natural Capital: what farmers and policy makers need to know. A report commissioned by the Food Farming and Countryside Commission and the Prince's Countryside Fund. Retrieved from London: https://ffcc.co.uk/publications/natural-capital-markets-report

Macroy, Richard. (2023, March). Breaking the Mould—Britain's New Office for Environmental Protection. *Journal of Environmental Law, 35*(1), 33–46. https://doi.org/10.1093/jel/eqac019

Martin, L., et al. (2020). Identity work in female-led creative businesses. *Gender, Work & Organization, 27*(3), 310–326.

Miraftab, F. (2009). Insurgent planning: Situating radical planning in the global south. *Planning Theory, 8*(1), 32–50.

Mittal, S. (2023). *The Wildlife Trusts urge Government to honour unfulfilled pledges for nature*. Retrieved from https://www.edie.net/the-wildlife-trusts-urge-government-to-honour-unfulfilled-pledges-for-nature/

Muir, K., et al. (2000). Shades of green in local authority policy-making: A regional case study. *Area, 32*(4), 369–382.

NALC. (2021). *What can local councils do on climate change*. National Association of Local Councils. https://www.nalc.gov.uk/library/our-work/climate-change/3598-what-can-local-councils-do-on-climate-change-2021/file

Nandy, L., et al. (2016). *The alternative: Towards a new progressive politics*. UK, Biteback.

NAO. (2014). *The impact of funding reductions on local authorities*. National Audit Office. https://www.nao.org.uk/wp-content/uploads/2014/11/Impact-of-funding-reductions-on-local-authorities.pdf

Newberry, D., & Dingemans, A. (2022). Levelling up and regeneration bill: Briefing note. Retrieved June, 17, 2023, from https://kennedyslaw.com/en/thought-leadership/article/levelling-up-and-regeneration-bill-briefing-note/

OBR. (2014). *Economic and fiscal outlook*. Retrieved from London. Previously available from https://cdn.obr.uk/37839-OBR-Cm-8820-accessible-web-v2.pdf

OEP. (2023). Government consistently failing to complete Post Implementation Reviews of environmental laws, finds OEP. Office of Environmental Protection. Retrieved from https://www.theoep.org.uk/report/government-consistently-failing-complete-post-implementation-reviews-environmental-laws

One Planet Council. (2022). Review of One Planet Development in Wales 2010–2021. September 2022. Retrieved from Wales: https://www.gov.wales/sites/default/files/publications/2023-05/atisn17403doc1.pdf

Our World in Data. (2022). Democracy in decline 1798–2022. https://ourworldindata.org/democracy

Owen, R., et al. (2022). SME Finance for Biodiversity ('FinBio') in the UK. Middlesex University. https://www.mdx.ac.uk/__data/assets/pdf_file/0034/675268/SME-Financing-for-Biodiversity-Project-Summary-MDX.pdf

Pew Research Centre. (2018). *Many across the globe are dissatisfied with how democracy is working*. Pew Research Centre. https://www.pewresearch.org/global/2019/04/29/many-across-the-globe-are-dissatisfied-with-how-democracy-is-working/

Routledge, P. (1996). Critical geographies and terrains of resistance. *Political Geography, 15*(6–7), 509–531.

Scottish Government. (2021). Scottish national just transition planning framework. Retrieved June 4, 2023, from https://www.gov.scot/publications/transition-fairer-greener-scotland/pages/5/

Shannon, J., & Burrowes, K. (2021). How governments can push towards a better tomorrow. Take on Tomorrow—Diversity and inclusion. Retrieved July, 9, 2021, from https://www.pwc.com/gx/en/issues/reinventing-the-future/take-on-tomorrow/government-challenges.html

Thornborough, J. (2021). *Trust in local government is still high, and policymakers should take advantage of it*. LSE. https://blogs.lse.ac.uk/covid19/2021/04/19/trust-in-local-government-is-still-high-and-policymakers-should-take-advantage-of-it/

Toledo, V. (2023). *Regenerative value networks: Principles and pillars of a new marketplace model*. Retrieved from LinkedIn https://www.linkedin.com/pulse/regenerative-value-networks-principles-pillars-new-model-toledo-1f/

Tollefson, J. (2021). COVID curbed carbon emissions in 2020 — but not by much. Nature Retrieved from https://www.nature.com/articles/d41586-021-00090-3

Tompt, J. (2016). *The meaning of Brexit? It's the movement. STIR to Action*. Dorset, STIR.

Treadwell, J., Tanner, W., Stanley, L., & Krasniqi, F. (2023). *Double Devo: Extending town and parish councils. The case for empowering neighbourhoods as well as regions*. Retrieved from London: https://www.uk.com/reports/double-devo-town-and-parish-councils/

Williams, J. (2013). The role of planning in delivering low-carbon urban infrastructure. *Environment and Planning B: Planning and Design*, *40*, 683–706.

Wills, J. (2016). *Locating localism: Statecraft, citizenship and democracy*. UK, Policy Press.

CHAPTER 3

The Politics of Planning in Delivering Sustainable Visions of Development

The Multi-faceted Role of the Planning Process

As leading UK directors of public health suggest in the foreword to *Getting research into practice: a resource for local authorities on planning healthier places*:

> *The environment we live in leads to and creates communities which, in turn, supports neighbourhoods that create a feeling of value, worth and social justice. To maintain and improve health and wellbeing, effective planning matters.* (PHE, 2019, p. 4)

Planning can shape the wider determinants of ill-health that lead to adverse health impacts (such as low levels of physical activity, obesity, air pollution, social isolation and even diet). It also shapes the provision of quality and safe green space, well-designed and insulated houses, proximity to employment, schools and other key services. The interaction of these elements can make the difference between households with quality or poor life choices over the span of their life-course. Incidences of deprivation, that is, how poor a given area is (normally measured at the lower super output area or ward level[1]) radically affects the opportunities available to people to get

[1] See https://geoportal.statistics.gov.uk/documents/d75d81ae855d470f8c2bb2425003457f/about

A. Burnett, *The Politics of Transition*, Palgrave Studies in Sub-National Governance,
https://doi.org/10.1007/978-3-031-40769-7_3

engaged in local placemaking: be that as stepping forward to being a local councillor, participating in local community engagement activities or having the confidence or self-belief that getting involved will make any difference.

The parameters through which the built and natural environment evolve are heavily shaped by the planning process and go on to have durable affects on people's well-being and their connection to place (Næss, 2016). Many studies have found that less well-off communities have higher rates of prescriptions (and dependency) and lower levels of Green Infrastructure (GI) provision, such as woodland, parks and gardens (Aerts et al., 2020; Douglas et al., 2018). The absence of this also serves to create a disconnect between the urban setting and nature, hampering the ability for society to understand its place within a wider ecosystem (and can account for why the climate and ecological emergency do not necessarily fall within the same breath, both in national policy terms and considered by some actors making climate and ecological emergency declarations).

In this book, we seek to uncover the politics of what happens when more radical visions of planning meet institutionalised practices and planning regimes in the context of English localism. Next, we start by setting out how the English planning system frames the discourse of sustainable development and how this in turn shapes the politics of purposively and effectively responding to climate and ecological emergencies and community resilience in a post-Brexit policy landscape.

I outline the recent planning reforms under the English localism policy agenda to embrace or reject emergent, alternative (claimed transformative) models for 'sustainable' development within a *relocalised* development model; that is, finding ways to build local resilience and encourage mutually reinforcing positive impacts that benefit local areas. We explore how this might be enabled under the localism agenda, introduced by the Conservative-Liberal Democrat Coalition Government (2010–2015) and formalised by the Localism Act 2011 in order to give greater autonomy to local actors, such as town and parish councils, residents and community groups. As former Prime Minister David Cameron suggested, *localism* and its overarching framing in the context of the 'Big Society' would "take power away from the elite and give it to the person on the street".

We will focus on planning's contribution to sustainable development alternatives at a local level within the (post?-)localist English planning policy context and what linkages 'planning arenas', such as neighbourhood plans (also known as Neighbourhood Development Plans, NDPs) have with broader placemaking and political arrangements. Notably, whether land-use and urban planning tools and practice can contribute to sustainability transitions. The introduction of a new spatial level—the 'neighbourhood'—may have affected the politics of sustainability transitions: how they are conceived, how ideas travel within the planning process, and their effect on wider socio-technical processes. NDPs *may* have indeed become an access point for redefining sustainable placemaking and in reconstituting identities to promote the goals of the New Economy. As we shall see, localism is also being used by communities as an attempt to solicit subversive promotion of the Commons and relocalised objectives. However, analysis of neighbourhood plans for sustainable outcomes is underdeveloped, although there has been valuable practitioner research carried out (CSE, 2020). To do so, we present the different forms of localism and critiques in the application of community rights, including neighbourhood planning, and question whether localism is enabling entry points for inciting development alternatives.

The English Planning System: A Mechanism to Plan for Climate and Ecological Emergencies?

In the UK, sustainability has been constructed through political discourse to have a particular relationship to planning, especially through EU legislation which has shaped expectations around environmental impact, such as Environmental Impact Assessments (Cowell, 2013). With Brexit, these relationships pose opportunities and risks in terms of offering a stable yet innovative environmental (planning) framework.

The English National Planning Policy Framework (NPPF, which sets out the country's overall framework for planning policy) clearly states that "[t]he purpose of the planning system is to contribute to the achievement of sustainable development" (DLUHC, 2023, para 7) and seeks to promote a 'three-legged stool' of sustainability: economic, social and environmental,[2] as defined in paragraph 8. However, I argue that the NPPF, and other definitions of sustainability upon these three pillars, neglects to make explicit these 'political' dimensions of sustainability in its definition—that is, the politics of crafting (normative, value-led) sustainability transitions (reflected in the intersection between the three pillars in Fig. 3.1).

For instance, the NPPF alludes to the 'right type' of land, the 'right' places' and the 'right' time for development and what constitutes 'opportunities for growth' (DLUHC, 2023, para 8a). However, these are not uncontentious issues. Moreover, whose definition of a "strong, vibrant

[2] A definition common to much sustainable development practice, influenced by the UN-commissioned Brundtland Report, also known as Our Common Future (1987).

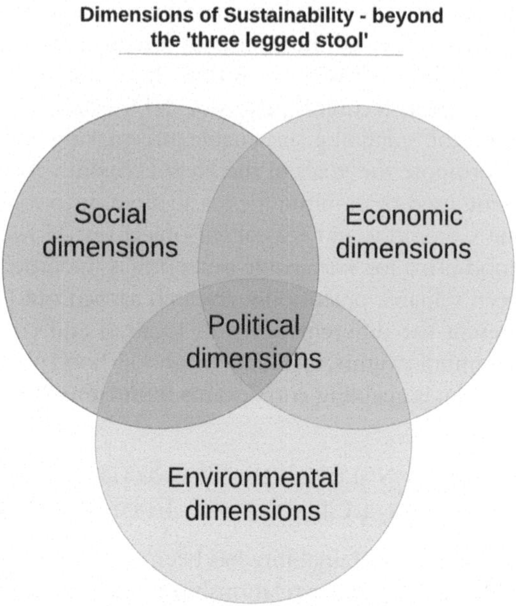

Fig. 3.1 Dimensions of sustainability beyond the 'three-legged stool'

and healthy community" (DLUHC, 2023, para 8b) is invoked is not a clear-cut response (Wills, 2012).

Deas has suggested that localism is an instance of "eroding [the nation state's] pre-eminence, and creating more space for non-state actors" (Deas, 2013, p. 9), whereby the state is broken down rather than progressively reformed. These reforms have been perceived as "permissive" legislation that anticipates that citizens will automatically step up (Wills, 2016a). Indeed, other authors have suggested that moves towards decentralisation are purposive attempts by regime actors to "rescale policy and decision-making processes to those scales where popular pressure can be muted or diffused and where state resources are insufficient to implement democratic decisions" (Peck & Tickell, 1994 in the words of Nicholls et al., 2013, p. 9). Similarly, Paddison argues that the consensual, deliberative model of localism is too fixated on process and focuses on *how* decisions on particular policy matters are made, rather than whether these issues should be on the agenda in the first instance (Paddison, 2009, in Deas, 2013).

For instance, the notion of development within urban planning is meant to reflect a symbiotic relationship among stakeholders and planning entities (issues and physical form to be moulded into place over time), yet in the NPFF these relationships have tended to be framed around the notion of growth as a driving force for development (Cobbinah & Darkwah, 2017). Similarly, Cowell has suggested that while the NPFF sets out the overall principles that underpin the planning process, including local plans and neighbourhood plans, it is a "device for meta-governance" that gives market actors undue preference due to localism's alignment to economic growth at the expense of effective regulatory instruments (Cowell, 2013). Felicetti suggests that localism, as proposed by the government, is couched in neo-liberal rather than *Transition* values (Felicetti, 2013)—see Table 4.1 Comparative analysis of Hopkins' assessment between localism and relocalisation and Frome's Independent for Frome's (IfF) independent political agenda.

Examples of this include that, in an earlier 2012 version of the NPPF growth is formalised within the discourse of sustainable development and strongly correlated to a *competitive* growth agenda:

> ***Development means growth.*** *We must accommodate the new ways by which we will earn our living in a* ***competitive*** *world...****So sustainable development is about positive growth****—making economic, environmental and social progress for this and future generations.* (DCLG, 2012, i, emphasis added)

There are also ten other mentions of 'growth' in the 2019 version of the NPFF and 12 in the December 2023 version.

A strong correlation between sustainable development and growth implies an ecological modernisation discourse (see Table 2.1) which emphasises economic growth without genuine attempts to encourage lower-carbon lifestyles (Hajer, 1997). Cowell argues that "(s)ince 2000, the environmental protection role of planning has been weakened in the UK as successive Westminster governments have 'streamlined' the system and reduced the scope for challenges to economic development on environmental grounds" (Cowell, 2016). Below, we outline key policy changes made by central government over the last decade and its consequence for net-zero planning at the local level.

Government Policy—Driving Deregulation and a Loosening of Net-Zero Standards?

While introducing new controls and powers at a local level, recent UK Conservative Governments have demonstrated what the Royal Institution of Chartered Surveyors (RICS) describes as a retreat from "prescriptive regulatory approaches", a move that has continued since the mid-1990s (RICS, 2016, p. 9). For instance, the removal of the Code for Sustainable Homes (CfSH)—partly because it was felt by the central government to encourage a 'box-ticking' approach where the overarching design and use of buildings became less important than the credits awarded for the inclusion of specific technologies (RICS, 2016).[3]

Some authors have suggested that these types of changes made with regard to housing energy efficiency standards encompass technical and environmental, rather than social dimensions (Pickvance, 2009). Moreover, commentators have criticised the government's current approach to low-carbon buildings for failing to recognise the role of strong regulation for encouraging innovation, notably in the removal of protections of conversation of employment to residential uses in some areas under deregulated Permitted Development (PD) which were more or less protected from development since 1948. Without wider regulation and structured policy goals, this can lead to too much emphasis placed on individual house-builders delivering ambitious net-zero buildings (RICS, 2016; Hodson & Marvin, 2013)

Table 3.1 lists some key policy changes and their implications for planning for net-zero and nature-positive development (Please note that while this list of policy changes is comprehensive, it is by no means exhaustive).

Despite the attempts by Government to pull the reigns on low-carbon planning in some of the measures described above, many communities under have sought to use the rights bestowed to them under the Localism Act to include local energy-related criteria in their progressing through neighbourhood plans. Indeed, in their 2017 *Guidebook to Low-Carbon Neighbourhood Planning*, the Centre for Sustainable Energy (CSE) later came to push the boundaries of what government policy allowed in terms of advancing localised low-carbon ambitions and suggested it is a *legal* duty for NDPs to address climate change (CSE, 2017). Thus, encouraging Neighbourhood Planning groups to be as ambitious as possible (which conflicts with some of the policy aims of encouraging standardisation in low-carbon building standards) and using the government's own

[3] While low-carbon homes halved in cost between 2011–2014, only 233 completed CfSH Code Six were completed by 2013 (2013 Element Energy Study in RICS, 2016). Higher-level CfSH codes (i.e., 5 and 6) were few and often tied to demonstration social housing projects awarded by the government's Homes and Community Agency (HCA) funds.

Table 3.1 Policy changes and their implications for planning for net-zero and nature-positive development

Time	*Policy/Legislation*	*Action*	*Effect*
1995	Town and Country Planning (General Permitted Development) Order	Introduced PD rights which allow certain types of development to be carried out without the need for a full planning application.	Weakened planning's control function where PD rights applied. NB: PD rights continued to be expanded in 2010, 2012, 2013, 2015, 2016, 2020, 2021, 2023 and 2024.
2008	Climate Change Act	Set legally binding carbon reduction targets to achieve an 80% reduction in emissions by 2050 (compared to 1990 baseline levels).	Intended to drive action to reduce emissions in different sectors, including development (though buildings continue to be one of the most inefficient areas of the UK economy).
2012	NPPF (paras 11–14)	Introduction of the presumption in favour of sustainable development; some exceptions to its application (i.e. where development conflicts with an up-to-date local plan, if the development is not deemed sustainable or in an Area of Outstanding Natural Beauty—since renamed National Landscapes).	Curtailment of the integrated role of planning and (environmental) sustainability. Gives undue favour to developers and encourages speculative development (see Wootton, 2017).
2013	Growth and Infrastructure Act 2013	Aimed to stimulate economic growth by streamlining the planning process for major infrastructure to support the housing market. It included provisions to expedite development projects, enhance compulsory purchase powers, and introduced a new system for planning obligations, ultimately aiming to boost investment and facilitate development in key sectors.	Mainly focused on major and infrastructure development and emphasised 'proportionality' in the documents submitted in support of a planning application; localism powers largely unaffected.

(*continued*)

Table 3.1 (continued)

Time	*Policy/Legislation*	*Action*	*Effect*
2014	Housing Standards Review	Effectively prevented any Neighbourhood Plan or Local Planning Authority (LPA) from setting additional local technical criteria (e.g. energy policies) in land-use policies.	Local Plans and Neighbourhood Plan low-carbon ambitions curtailed, particularly relating to on-shore wind.
2014	Community Energy Strategy	Aimed to support the development of community-led renewable energy projects and empower local communities to participate in the energy transition.	Aimed to support the community energy sector, but hampered by the cuts to the Feed-in-Tariff (FiT), which made some new projects unviable and returns on investment.
2015	Zero-Carbon Homes (ZCH) and Allowable Solutions (withdrawal of)	A Labour government policy that aimed to meet net-zero by 2026. Was scrapped soon after the Conservatives won a majority in May 2015, who argued the ZCH target was not necessary as reforms would be tied to the updated Part L of the Buildings Regulations (though less stringent than that proposed by the ZCH policy).[a]	Eliminated the requirement for new residential buildings to meet specific carbon reduction standards, arguably setting back UK decarbonisation in the construction and development sector.

(*continued*)

Table 3.1 (continued)

Time	*Policy/Legislation*	*Action*	*Effect*
2015	Deregulation Act	Formally withdrew the Code for Sustainable Homes (CfSH) as a voluntary low-carbon buildings framework, thus removing the ability for local authorities (and NDPs) to set targets for energy-efficient dwellings.[b] Extended PD, particularly around conversion of employment to residential in some areas; removal of internal space standards.	Absence of a common framework for low-carbon design and construction (RICS, 2016). Withdrawal of a policy framework for several areas that were covered in the scheme, including ecology, and the sustainability and embodied energy of building materials (see BRE, 2015). LPAs could not uphold environmental and design standards through the planning system, including enforcing sustainable locations for development (Remøy & Street, 2018). Inhabitants likely to face impending reproofing costs and materials. Freedom for developers to build houses with no windows or be suitably insulated (where PD rights apply). Undermines the legislative ambition to enhance liveability and well-being in the Homes (Fitness for Human Habitation) Act 2018.
2017	Neighbourhood Planning Act	Aimed to strengthen Neighbourhood Plans and make the planning process more reflexive, including a revision of both Local Plans and Neighbourhood Plans at least once every five years.	Greater weighting of a post-Examination Neighbourhood Plan prior to it being made. Local Planning Authorities must inform town and parish councils of any planning applications being put forward within a designated neighbourhood area.

(*continued*)

Table 3.1 (continued)

Time	*Policy/Legislation*	*Action*	*Effect*
2018	NPPF (short-lived version)	Included provisions for encouraging development on smaller sites; expanded the definition of affordable housing; greater emphasis on design codes; encouraging LPAs to use compulsory purchase powers to bring forward sites and setting out a standard methodology for calculating housing need. Permitted future investment in coal. NB: the government did not apply a Strategic Environmental Assessment (SEA) to its own policy-making.	Friends of the Earth (FoE) criticised the government for allowing, perceived as a "frackers' charter" and failure to hold up standards (see Laker, 2018).
2018	25-Year Environment Plan (25-YEP)	To provide a long-term vision and strategic approach to improving the environment, focusing on areas such as air and water quality, biodiversity, waste reduction and climate change mitigation.	Set a framework for subsequent policy and investment in environmental initiatives.
2019	Amendment to the Climate Change Act	Amended the legally binding carbon reduction target to achieve net-zero by 2050.	Led some (i.e. CSE and the TCPA, 2023) to encourage planning policy to see the need to achieve net-zero as a legal duty. However, conflict in other policies meant uncertainty in how and whether plans can be this ambitious.
2019	Future Homes Standard	England-wide standard aimed at tackling the climate emergency. Will come into effect from 2025 to ensure new homes produce 70–80% less carbon emissions. Complementing the updated Part L of the Building Regulations.	Unclear measures if not compliant. Reliant on wider low-carbon infrastructure to be optimised concurrently.

(*continued*)

Table 3.1 (continued)

Time	*Policy/Legislation*	*Action*	*Effect*
2021	Heat and Buildings Strategy	Sets out measures to decarbonise homes, commercial, industrial and public buildings by 2050.	Critiqued for not emphasising the urgency to reduce embodied carbon emissions from UK buildings through materials.
2021	Environment Act	Introduction of Biodiversity Net Gain (BNG)—a mandatory commitment for development to contribute to at least a 10% improvement in baseline ecological assessments (which mostly came into force in 2024). Formalised role for Local Nature Partnerships (LNPs) to deliver on local nature recovery strategies.	Helped to centre stage nature into planning instruments and give rise to new forms of environmental governance to achieve these ambitions. New markets for biodiversity offsetting are also emerging—though, hampered by the lack of formal regulation in this area—which are seeing marketisation of nature as a tradable commodity, building on previous Payments for Environmental Services (PES) schemes.
2023	Environmental Improvement Plan (EIP) and Associated Outcome Indicator Framework (OIF)	First revision of the 25-Year Environment Plan (25-YEP). Sets out how the government will engage with landowners, communities and businesses to deliver environmental objectives and the interim targets to measure progress towards this. The OIF describes environmental change that relates to the ten goals in the 25-YEP.	Integrates a nature-positive approach to planning. Replaces a lacuna in a national environmental monitoring framework (after the removal of the national indicators in 2015).
2023	Levelling-Up and Regeneration Act	Sets out that the environmental effects of plans and projects will be assessed through the new Environmental Outcome Reports, rather than SEAs or EIAs. Amends the Countryside and Rights of Way Act 2000 to include a duty to further the aims of protected landscapes, rather than have due regard to them.	Indicates increasing separation from EU policy in a post-Brexit environmental policy landscape. Emphasis on just, redistributive transitions and the role of planning to support this among other mechanisms. Duty to support protected landscapes extends beyond boundaries of where they are geographically situated.

(*continued*)

Table 3.1 (continued)

Time	*Policy/Legislation*	*Action*	*Effect*
2023	Revision of the National Planning Policy Framework	The influence of Neighbourhood Plans was strengthened to uphold policies in favour of a plan that was made less than five years ago (before it was two years). LPAs have more remit to deviate from the national standard method for calculating housing need. Removed the need to review green belt boundaries during reviews of local plans.	Likely means that speculative housing planning applications will be less likely to be approved where there is a conflict with a Neighbourhood Plan. Gave LPAs more flexibility to calculate housing need. Strengthened green belt protections, likely resulting in less housing pressures in these areas. The relaxation of housing supply only applies if an up-to-date local plan is in place, or an emerging one is suitably advanced; many LPAs are not in this position, so in practice there may be limited impact on housing pressures.

[a] See https://www.gov.uk/government/publications/conservation-of-fuel-and-power-approved-document-l (last accessed 12.3.19)

[b] Since the Building Research Establishment Environmental Assessment Method (BREEAM) was not adopted as a government voluntary standard it was not affected by these changes in the commercial sector

international climate policy commitments and national targets to elicit (local) low-carbon policies.[4]

Nevertheless, critics have suggested that the UK government does not go far enough to ensure that planning delivers on the net-zero targets set out in the 2008 Climate Change Act (as amended). For instance, the Town and Country Planning Association (TCPA) has suggested that the current planning system lacks long-term strategic planning, leading to fragmented and unsustainable development. The TCPA also highlights the need for affordable housing provision and criticises the focus on short-term economic gains over social and environmental objectives. Additionally, they advocate for a more inclusive and participatory planning process,

[4] Indeed, such guidance was inspired by an Examiner who suggested that if NDPs allocate additional housing sites that go beyond any local targets set by an LPA or its five-year land supply of deliverable sites then additional technical standards were construed at the time as possible on these sites. See: https://www.gov.uk/guidance/housing-and-economic-land-availability-assessment (last accessed 23.1.19).

aiming to give communities a stronger voice in decision-making and ensuring that planning serves the public interest.

The Architects Climate Action Network has called for more emphasis on enforcement of the energy aspect of building regulations, which is currently poor, and for the regulations to cover the whole-life carbon emissions of buildings. The CCC has suggested that the current Energy Performance Certificate (EPC) rating system is inadequate. The Green Finance Initiative (GFI, 2021) has suggested that green passports should be introduced—digital passports that can provide guidance on actions required to improve buildings energy efficiency and comfort. They suggest these could be scaled up to include recommendations on low-carbon heat, indoor air quality, flood resilience, water efficiency and overheating and provide for a comprehensive plan to retrofit all properties to respond to the challenge of net-zero; green passports could be transferable between occupiers to support longer-term decarbonisation of properties and digitally capture EPC data which could be compared against other data over time.

Neighbourhood Plans—A Vehicle for Localised Net-Zero Configurations?

As embodied within the Localism Act (2011), a series of rights were conferred to local actors to actively contribute to placemaking. Notably, these included *NDPs, the Community Right to Bid, the Community Right to Build* as well as an attempt to bestow greater power to local councils under the *General Power of Competence*[5]—all underpinned by the NPPF (since DCLG, 2012).

NDPs were introduced as a right for parish and town councils (or in cities, Neighbourhood Development Forums) to develop a plan for their 'neighbourhood area'[6] that sets out local priorities on land-use and development. NDPs set out local development priorities and can add local flavour to existing strategic planning policies set by the LPA (such as district and unitary councils).

As this book will explore, NDPs play a clear role in understanding how existing institutional, spatial, social and political "configurations in urban and regional contexts are able to absorb, resist or reshape the dominant pathways" (Hodson & Marvin, 2013, p. 69). NDPs are a critical mechanism to consider within a broader placemaking strategy because they can indicate the extent to which ideas are contained, constrained, loosened or dispelled by the plan-making process.

[5] For an overview of Localism see Wills (2016b) and Ricketts and Field (2012).

[6] An area defined through formal consultation at the beginning of the NDP process.

One of the key drivers in the government's creation of NDPs was an expectation that by developing a plan, the community would in turn be more supportive of additional development in their areas if they felt more in control of development (see DCLG, 2012). However, there are some restrictions on their use and remit for local placemaking autonomy:

Box 3.1 The Steps of the Neighbourhood Planning Process

The NDP process

NDPs have distinct phases that they must go through according to the Neighbourhood Planning (2012) Regulations:

(i) Establishment of a steering group or Neighbourhood Development Forum, with a Link Officer (from the LPA) in an advisory capacity.
(ii) Approval of the neighbourhood area (by a six-week consultation led by the LPA).
(iii) Ongoing community consultation to support objectives and policies (evidenced in a Consultation Statement).
(iv) Collation of an evidence base to inform policies, including a Habitats Regulation Screening to determine if the plan will impact on protected (EU) habitats (and an Appropriate Assessment if it does); and a Sustainability Appraisal, where significant impacts on protected habitats and species are considered likely (note that some changes to post-Brexit environmental policy will affect which environmental compliance documents are required in the future, i.e. SEAs are now being replaced by Environmental Outcomes Reports, see Table 3.1). Compliance with national and local planning policy needs to be set out in a Basic Conditions Statement.
(v) Regulation 14 a local (minimum six-week consultation) on the Pre-Submission NDP, leading to a revised Plan arising from comments received during the consultation at this stage.
(vi) Regulation 16 on the amended plan 'Submission' version.
(vii) Formal Examination by an independent examiner.
(viii) Modifications on the Reg 16 (Submission) Version, unless it fails examination.
(ix) Referendum on the Submission Plan.
(x) Adoption of the 'made' plan.
(xi) Review/modification of a made plan.

- NDPs are only concerned with development and use of land in line with Section 38A of the Planning and Compulsory Purchase Act (HMG, 2004).
- Desirable community projects do not form part of a plan, rather these form non-plan related 'action points' that are appended to an NDP.
- An NDP can only be carried out by a qualifying body, which must be an existing local government where this exists (i.e., parish or town councils), or (in the case of a city) a neighbourhood forum, which can be established with 21 members of local community groups.
- NDPs themselves must be in 'general conformity' to the Local Plan (or Core Strategy) set by the LPA and other national and (previously) EU legislation (though with the passing of the Levelling Up and Regeneration Act which came into force in October 2023, EU legislisation is becoming superceded - such as the replacement of SEAs by new Environmental Outcome Reports (see Table 3.1).

Once 'made' an NDP becomes part of the local planning framework—the set of policies, guidelines and regulations that govern land use and development within a specific local area—together with other key planning documents (including a Local Plan)—to determine decisions on planning applications made by the LPA.

As of March 2024, to date over 3000 community groups and local councils have initiated the NDP process, with over 1215 made plans—illustrated by data provided by the Department for Levelling Up Housing and Communities DLUHC (see Table 3.2).

Table 3.2 Number of Neighbourhood Plans at each stage of the NDP process

NDP Stage	*Number at Each Stage*
0. Plan Made	1215
1. Passed Referendum	453
1.5. Failed Referendum	11
2. Referendum	12
3. Passed Examination	21
3.5. Failed Examination	11
4. Submitted for Examination	21
5. Pre-submission Consultation	75
6. Designated	1248
Total	3049

Source: Locality/DLUHC (2024)

DLUHC data summarised in the above table reveals that the two LPAs with the most made plans so far are Cornwall Council and Wiltshire Council (35). Those with the least number of applications include some city councils, a few London boroughs and more sparsely populated national park authorities such as the Yorkshire Dales or Exmoor (and in quite a few of these cases, local planning authorities have only had one application to date) (DLUHC, 2024). Though this is not necessarily surprising given there are generally fewer potential neighbourhood areas to designate in this areas, comparatively. For a helpful map of the concentration of applications to become a neighbourhood area (see Planning Resource, 2023). The map below shows the geographical coverage of neighbourhood planning groups at the end of 2022 (Fig. 3.2).

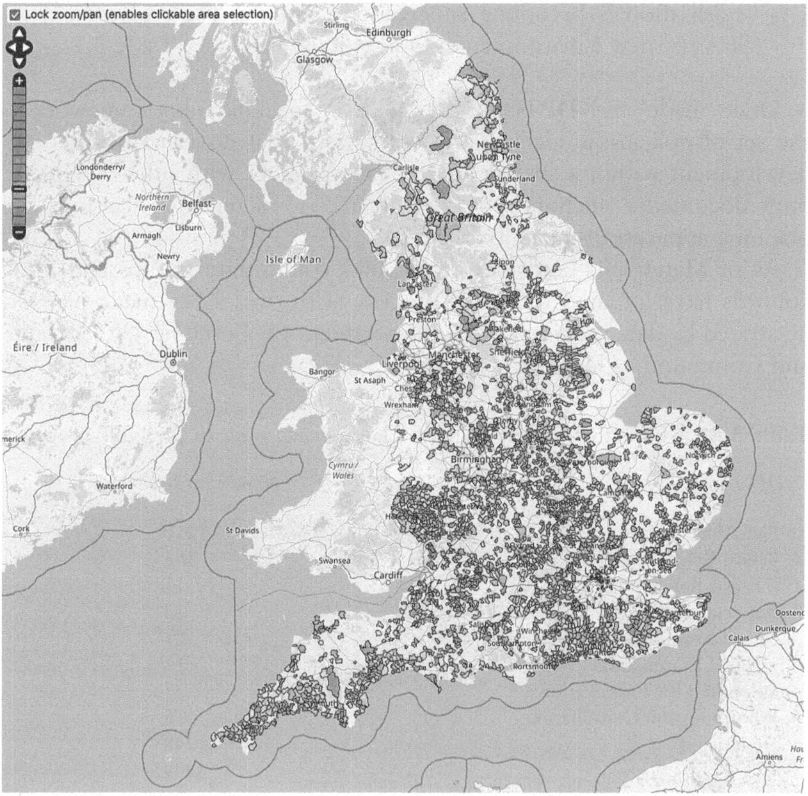

Fig. 3.2 Coverage of neighbourhood plans across England (LG Inform, LGA, 11.11.22)

The Scope of Low-Carbon Localism

Various national policy changes outlined in Table 3.1 have therefore both curtailed and enabled community influence on (low-carbon/net-zero) planning policy-making, as indicated in the Table 3.3 below. The following table sets out some examples of how NDPs are used to advance low-carbon (or relocalised) development. Indeed, since the policies of previous Conservative governments, there is much more encouragement by local authorities to use NDPs as a tool to advance sustainability objectives. For instance, Wiltshire Council Reg 14 Climate Change Guidance states that NDPs should be "actively encouraged to have the mitigation and adaption to climate change as their golden thread" (Wiltshire Council, 2023, p. 2).

Table 3.3 Key policy changes since the introduction of NDPs that affect low-carbon ambition and delivery

Area	*Example of potential influence*
Low-carbon ambitions (general)	Aspiration to be net-zero and respond to local authority climate emergency declarations, for example, by 2030 (earlier than the statutory 2050 set out in the 2008 Climate Change Act (as amended), which was passed in 2019). Indicate support for community energy or intention to create a group. To encourage BREEAM Excellent[a] in commercial development (if not already covered in the Local Plan) or the use of BREEAM Communities framework. Adhere to One Planet Living (OPL) principles or the Sustainable Development Goals (SDGs).
Energy (renewables)[b]	Supporting community energy schemes or the provision of stand-alone renewable energy sites. Specific on-site renewable energy generation or stand-alone solar ground mount sites, including wind energy. District heating schemes. Targets for on-site renewables generation in new development.
Energy efficiency	Upgrading existing dwellings. Policies to support retrofitting, including responsible retrofitting of historic / traditional buildings (see Historic England, 2023).
Sustainable Transport	Encouraging active travel. Criteria for new builds to be walking distance from the town centre or linking recreational and residential areas. Encouraging cycle pathways. Re-ordering priorities between cars and pedestrians (with the latter taking precedence).

(*continued*)

Table 3.3 (continued)

Area	*Example of potential influence*
Biodiversity	Protection of ecosystems through wildlife corridors (mapping natural assets to strengthen links in ecosystem chains). New development can integrate with existing green infrastructure. Contributing to biodiversity net gain (BNG—a minimum of 10% is set out in the Environment Act, with some LPAs aiming even higher, for example, in some areas in Surrey and the emerging Wiltshire Council Local Plan which set a 20% BNG target). Habitat creation, woodlands and wetlands. Local Green Space (LGS) designation.
Water	Encourage the use of Sustainable Urban Drainage (SUDs), incorporated into landscaping with biodiversity benefits. Greywater recycling through design features. Supporting measures to adapt to flooding and flood resilience—including siting development away from main areas of flood risk (adding local measures to implement the NPPF's sequential test and exceptions test to direct development away from areas of high flood risk and ensure appropriate mitigation measures for development in a flood-prone area).
Self-build	Opportunity for Neighbourhood Plans to identify self-build plots. Potential for highly sustainable, lower-cost housing (outside of the stringent affordable housing definition in the NPPF).
Local Employment	Provision of start-up office facilities / micro-business hubs in rural settlements, localising working.
CIL spending	Bus interchange, cycle routes. Renewables could also be funded through the Community Infrastructure Levy.

[a]OPL, community energy and BREEAM were all previously proposed as policies in earlier versions of Frome's Neighbourhood Plan.

[b]Some NDPs were hoping to set additional measures to secure reduced emissions through local policies through the CfSH. However, NDP groups, such as Frome, were forced to remove these policies after the Housing Standards Review.

Many of these measures can indirectly help support local businesses, particularly small and medium-sized enterprises (SMEs) to provide net-zero retrofitting services and aid greener business models within communities. Achieving net-zero design also leads to various co-benefits such as reducing energy bills and enhancing the liveability of new and existing homes.

However, net-zero design may well lead to a change in style and the use of materials as the built environment increasingly utilises different building fabric - such as modular designs as well as sustainable materials such as hemp and straw - and this may impact on a sense of place if traditional styles are deviated from. Again, reverting to Wiltshire Council's Neighbourhood Planning Climate Change Guidance (p.2), the Council declares that "A new carbon neutral vernacular is needed" but in such a way that these protect and enhance local heritage assets.

Engaging Environmental Issues in NDPs—Potential and Constraints

In a joint paper published as part of early informational interviews for this research (see Bradley et al., 2017), I found that representatives from environmental organisations, such as Transition Initiatives and Friends of the Earth (FoE) had begun to engage in NDPs to advance alternatives to growth-dominated discourses. Transition Initiatives were using their own organisational documents to influence the NDP evidence base and policies, such as using a Transition Initiative-created *Energy Descent Plan* or a sustainable transport strategy. Moreover, where Green Party councillors were engaged (such as town, district or county) this enabled formal links between the NDP, the Transition Initiative and elected (green) actors. In one case *Transition* could thus drive its own agenda forward through the NDP process, particularly since the sustainability working group of the NDP was co-opted into the Transition Initiative itself.

A key challenge for 'neighbourhood planners' actors is developing a plan that is ambitious and at the same time implementable. I found as part of the formal governance of the NDP groups, some environmentalists took care to avoid radical sustainability discourse, while other strategies included leaving some more contentious issues (such as on-shore wind) off the table, inviting sympathetic local actors and organisations to help amplify the sustainability message or enrolling national actors, such as such as CSE (who offered a low-carbon neighbourhood planning support scheme) to validate their low-carbon concerns.

Some commentators have been overtly critical of neighbourhood plans, particularly around the mismatch of the early ambitions of government discourse to empower communities with some stark realities on their actual scope of influence. Parker et al., even suggest, "our view is that innovation is being constrained if not entirely suppressed" (Parker et al., 2017, p. 446), for these authors, the highly prescriptive nature of the NDP process requires "pliant communities"—who act as the "good citizen" and abide by "technologies of power" through which the government's planning intentions (e.g., to build more houses) are channelled (Parker et al., 2017, pp. 448–449).

For instance, Parker and Slater (2017) found that of 130 NDPs that had passed a referendum by Spring 2016, almost half required major modifications required by the Examiner. Such changes may be indicative of a 'rescripting' community intentions (Parker et al., 2015; Parker & Street, 2015; Parker et al., 2017). They also suggest that outcomes are necessarily bounded; the governance of Neighbourhood potential Plans rests on prescriptive, overtly technical and complex parameters. Felicetti also argues that an often bi-lateral, limited engagement of Transition Initiatives with formal government institutions in the context of localism has only served to formalise exclusionary tendencies, rather than promoting grassroots activism (Felicetti, 2013).

Wills suggests that the current local infrastructure for engagement in the polity is a "weakness" in fulfilling the promise of localism and there are significant inequalities in the geographic distribution of power and a community's ability to respond (Wills, 2016a, p. 1). Fundamentally, while NDPs seek to broaden engagement and influence placemaking there appears to be a *spatial fragmentation* of localism within the NDP process (McGuinness & Mawson, 2017), raising questions over its legitimacy as an emancipatory planning instrument. For instance, earlier data suggests that NDPs had an average 89% in favour of the plans (MHCLG, 2016) but only 32.4% average turnout (total electorate 335,000)—as of August 2016 (see Carpenter, 2016). In addition, the think tank Res Publica found that ten local authorities with the most deprived areas in England also had the fewest designated NDPs, and the opposite was true for more affluent areas (see Sharman, 2016).

Not recognising power and inequality of representation in NDPs means that "austerity localism envisions decentralizing power to certain local people"—i.e., the 'default actors' with existing resources and expertise (Featherstone et al., 2012, p. 178). This may risk deepening inequalities within and between places. Despite more recent efforts by LPAs to deliver climate change guidance to NDPs since the onset of 'climate emergency declarations', there is also a danger that there is a lack of skills and capacity for those expected to deliver localism outcomes to effectively respond to the challenges of 'sustainable development' (Fiona Reynolds, cited in Ricketts & Field, 2012; Wills, 2016a).

More broadly, there is still a long way to go before the planning process becomes more adaptive to respond to the net-zero and nature-positive challenges (and opportunities), such as addressing the depletion of biodiversity while reducing carbon emissions. The TCPA-commissioned the Raynsford Report to investigate the current state of the UK planning system and whether it is fit for purpose to respond to multiple environmental challenges, such as flooding, biodiversity loss, sustainable transport and the lack of affordable housing. This report indicates that "there is a real appetite for a revitalised planning system capable of effective and visionary place-making … organised to deliver solutions to the big challenges we will increasingly face" (TCPA, 2018, p. 3). It calls for "new and improved guiding principles, structures, relationships and processes to the planning system with the potential to deliver real economic, social and environmental advances" (ibid., p. 3).

However, there is also potential to celebrate localism and what it can enable. Wills suggests that "localism represents a challenge to the model of government and politics that developed during the twentieth century and in theory at least, it marks a new phase of statecraft" (Wills 2016a, p. 44). For localism to be effective, Wills suggests both new civic infrastructure and increased capacity for people to engage in public life are required to create "new political ecosystems"; failure to do so may risk a transfer of power to a new set of elites empowered through government localism reforms. Such an approach should reflect a move towards more horizontal approaches that younger people are accustomed to for a more 'networked' approach to technological interaction (Wills 2016b, p. 19). Indeed, Wills also suggests that NDPs may yet encourage a new form of politics that is not yet present in academic research (Wills, 2016a).

Alternative Political Ecosystems as Delivery Nodes of a More Radical Development

More than a decade ago, Hodson and Marvin suggested that a viable alternative to neo-liberalism is not yet formed: any such alternative would need to see "new power relationships coalesce to develop a view of state-space that seeks to redistribute state resources, forms of knowledge and categorisation in ways that seek to reinvigorate places and regions" (Hodson & Marvin, 2013, p. 29). This does not mean to say that such endeavours have not progressed significantly since climate emergency discourse framed a coalescence of climate resilience and community-based planning—both on paper and increasingly in practice.

Parker et al., draw on May (2008) to suggest that NDPs remain a "foot in the door" for a "hope" towards more progressive local politics (Parker et al., 2017). They propose that social and spatial belonging to particular places can be used to create oppositional identities that encourage antagonism against the seeming governmentality of the NDP and broader planning processes.

For instance, co-founder of the *Transition* movement, Rob Hopkins, suggests that government reforms have actively encouraged Transition Initiatives to link to broader mainstream processes of government design and, in effect, for Transition Initiatives to become developers (Hopkins, 2014a, 2014b). Hopkins suggests *Transition* utilises community-based planning mechanisms such as Community Land Trusts (CLTs), and Community Right to Build Orders (CRtBOs) to influence development outcomes through community ownership of assets (Hopkins, 2015). Hopkins has also proposed invoking discursive counter-arguments, for instance, turning NIMBY (Not In My BackYard) from a perception of resistant, progress-hampering residents to SWIMBY (Something Wonderful In My BackYard).[7] Hopkins also believes that Transition Initiatives should also actively withdraw their consent for proposed applications that are seen to be unsustainable (Hopkins, 2015). The practices of the *Transition* movement also resonate with shared, community-managed assets and cohousing (intentional communities run by their residents with a mix of self-contained and shared community spaces)[8] which could become viable alternatives supported by or channelled through localism rights and NDPs.

Conclusion

In the chapters that follow, we will be exploring how NDPs are, in practice, helping to reconfigure local contexts to create new (and implementable) visions of development that reflects genuine input from communities in shaping these objectives. Importantly, we also assess how the local political context helps to shape the opening of these opportunities by closely examining how an English market town in Frome, Somerset set the scene for a pioneering town council and active environmental actors in the town as they sought to 'take control' of local planning and placemaking to achieve

[7] See Hopkins (2014b) for a refreshing SWIMBY manifesto.

[8] Cohousing originated in Scandinavia in the 1970s. There are 19 operational cohousing schemes in the UK, notably Lilac in Leeds, Stroud/Bicester and 60 groups actively developing projects. For more information see https://cohousing.org.uk/ (last accessed 3.12.23).

sustainable and climate resilience objectives. We examine the intersection between localism, sustainability 'transitions', placemaking and politics and consider what this innovative case can tell us about the scope for grassroots innovation in planning for climate and ecological emergencies and the conditions that are necessary for such action to gain momentum, both locally and beyond. In particular, the route into formal local government administration and regimes by environmental movements and groups, such as political representation in local government or by contributing to the plan-making process, are vital clues into its transformative potential and the limits, and opportunities, of social movement-directed change through formal and informal processes. We will go on to reflect on whether the planning process can live up to its claimed refashioned role of enabling bottom-up voices to have a greater influence on the nature of placemaking; and if not, what other strategies can lead to possible place-based, regenerative transition pathways?

References

Aerts, R., Nemery, B., Bauwelinck, M., Trabelsi, S., Deboosere, P., Van Nieuwenhuyse, A., et al. (2020). Residential green space, air pollution, socio-economic deprivation and cardiovascular medication sales in Belgium: A nationwide ecological study. *Science of The Total Environment, 712*, 136426. https://doi.org/10.1016/j.scitotenv.2019.136426

Bradley, Q., Burnett, A., & Sparling, W. (2017). Neighbourhood planning and the spatial practices of localism. In S. Brownhill & Q. Bradley (Eds.), *Localism and neighbourhood planning. Power to the people?* Policy Press.

BRE. (2015). Housing Standards Review. (Last accessed 15.5.17). Retrieved from https://www.bre.co.uk/housing-standards-review

Carpenter, J. (2016). Analysis finds average neighbourhood plan turnout of 32.4%. https://www.planningresource.co.uk/article/1406239/analysis-finds-average-neighbourhood-plan-turnout-324

Cobbinah, P. B., & Darkwah, R. M. (2017). Urban planning and politics in Ghana. *GeoJournal, 82*, 1229–1245. https://doi.org/10.1007/s10708-016-9750-y

Commission, T. B. (1987). *Our common future*. Oxford University Press.

Cowell, R. (2013). The greenest government ever? Planning and sustainability in England after the May 2010 elections. *Planning Practice & Research, 28*(1), 27–44.

Cowell, R. (2016). *Planning and the environment after the EU referendum*. https://plannersnetworkuk.wordpress.com/2016/04/29/planning-and-the-environment-after-the-eu-referendum/

CSE. (2017). *A Guidebook for low carbon neighbourhood planning* (3rd ed.). Retrieved from Bristol, UK: https://centreforsustainableenergy.ams3.digitaloceanspaces.com/wp-content/uploads/2023/03/18215641/neighbourhood-planning-in-a-climate-emergency-feb-2020.pdf

CSE. (2020). *A guide to policy writing and community engagement for low-carbon neighbourhood plans.* Retrieved from Bristol, UK: https://centreforsustainableenergy.ams3.digitaloceanspaces.com/wp-content/uploads/2023/03/18215641/neighbourhood-planning-in-a-climate-emergency-feb-2020.pdf

CSE, & TCPA. (2023). *Spatial planning for climate resilience and Net Zero.* Retrieved from London: https://www.theccc.org.uk/publication/spatial-planning-for-climate-resilience-and-net-zero-cse-tcpa/

DCLG. (2012). *National planning policy framework.* DCLG.

Deas, I. (2013). Towards post-political consensus in urban policy? Localism and the emerging agenda for regeneration under the Cameron government. *Planning Practice & Research, 28*(1), 65–82. https://doi.org/10.1080/02697459.2012.699228

DLUHC. (2024). Key neighbourhood planning data. 22nd March 2024 (accessed 17th April 2024). https://neighbourhoodplanning.org/toolkits-and-guidance/key-neighbourhood-planning-data/

DLUHC. (2023). *National planning policy framework.* London: HMG Retrieved from https://assets.publishing.service.gov.uk/government/uploads/system/uploads/attachment_data/file/1182995/NPPF_Sept_23.pdf

Douglas, O., Scott, M., & Lennon, M. (2018). *Greenspace and urban greening benefits for health and well-being across the life-course—A pathway for the operationalisation of the green infrastructure approach.* Routledge.

England, H. (2023). *Retrofit and energy efficiency in historic buildings.* https://historicengland.org.uk/advice/technical-advice/retrofit-and-energy-efficiency-in-historic-buildings/

Featherstone, D., Ince, A., Mackinnon, D., Strauss, K., & Cumbers, A. (2012). Progressive localism and the construction of political alternatives. *Transactions of the Institute of British Geographers, 37*(2), 177–182. https://doi.org/10.1111/j.1475-5661.2011.00493.x

Felicetti, A. (2013). Localism and the transition movement. *Policy Studies, 34*(5–6), 559–574. https://doi.org/10.1080/01442872.2013.862449

GFI. (2021). *Building renovation passports: Creating the pathway to zero carbon home.* Retrieved from London: https://www.greenfinanceinstitute.co.uk/wp-content/uploads/2021/03/GREEN-FINANCE-BUILDING-RENOVATION-final.pdf

Hajer, M. A. (1997). *The politics of environmental discourse: Ecological modernization and the policy process.* Oxford University Press.

HMG. (2004). Planning and Compulsory Purchase Act.

Hodson, M., & Marvin, S. (2013). *Low carbon nation?* Routledge.

Hopkins, R. (2010). Localisation and resilience at the local level: The case of Transition Town Totnes. (PhD). University of Plymouth, Plymouth. https://pearl.plymouth.ac.uk/handle/10026.1/299

Hopkins, R. (2014a). Our month on rethinking real estate: Why I'm proud to be a SWIMBY. https://www.transitionnetwork.org/blogs/rob-hopkins/2014-10/our-month-rethinking-real-estate-why-i-m-proud-be-swimby

Hopkins, R. (2014b). *Rob Hopkins speaking to at an event in Salisbury, Wiltshire to support the transition movement, locally.*

Hopkins, R. (2015). Planning: The untapped power of community consent. https://www.transitionnetwork.org/blogs/rob-hopkins/2015-02/planning-untapped-power-community-consent

Laker, L. (2018). Government faces court action over 'illegal' planning policy. *The Guardian.* https://www.theguardian.com/environment/2018/sep/04/government-faces-court-action-over-illegal-planning-policy

May, T. (2008). *The political thought of Jacques Rancière: Creating equality.* Edinburgh University Press.

McGuinness, D., & Mawson, J. (2017). The rescaling of sub-national planning: Can localism resolve England's spatial planning conundrum? *Town Planning Review, 83*(3), 283–303.

MHCLG. (2016). New landmark with 200 communities now approving Neighbourhood Plans [Press release]. Retrieved from https://www.gov.uk/government/news/new-landmark-with-200-communities-now-approving-neighbourhood-plans

Næss, P. (2016). Built environment, causality and urban planning. *Planning Theory & Practice, 17*(1), 52–71. https://doi.org/10.1080/14649357.2015.1127994

Nicholls, W., Miller, B., & Beaumont, J. (2013). Introduction: Conceptualizing the spatialities of social movements. In W. Nicholls, B. Miller, & J. Beautmont (Eds.), *Spaces of contention: Spatialities and social movements* (pp. 1–23). Routledge.

Parker, G., Lynn, T., & Wargent, M. (2017). Contestation and conservatism in neighbourhood planning in England: Reconciling agonism and collaboration? *Planning Theory and Practice, 18*(3), 446–465.

Parker, G., & Slater, K. (2017). Taking stock of neighbourhood planning in England 2011–2016. *Planning Practice & Research, 32*(4), 478–490.

Parker, G., & Street, E. (2015). Planning at the neighbourhood scale: Localism, dialogic politics, and the modulation of community action. *Environment and Planning C: Government and Policy, 33*, 794–810. https://doi.org/10.1068/c1363

Parker, G., Lynn, T., & Wargent, M. (2015). Sticking to the script? The co-production of Neighbourhood Planning in England. *Town Planning Review, 86*(5), 519–536. https://doi.org/10.1080/14649357.2017.1316514

Peck, J., & Tickell, A. (1994). Searching for a new institutional fix: The after-fordist crisis and the global-local disorder. In A. Amin (Ed.), *Post-fordism* (pp. 280–315). Blackwell. https://doi.org/10.1002/9780470712726.ch9

PHE (2019). *Getting research into practice: A resource for local authorities on planning healthier places.* Public Health England (a PHE publication based on research by the Town and Country Planning Association and the University of West England) https://assets.publishing.service.gov.uk/government/uploads/system/uploads/attachment_data/file/951310/GRIP2_PHE_national_resources_151220_for_Gateway_2.pdf

Pickvance, C. (2009). The construction of UK sustainable housing policy and the role of pressure groups. *Local Environment, 14*(4), 329–345. https://doi.org/10.1080/13549830902764712

Planning Resource (2023) Mapped: neighbourhood plans (updated 14/04/2023) - accessed 3rd December 2023. https://www.planningresource.co.uk/article/1212813/mapped-neighbourhood-plans-updated-14-04-2023

Remøy, H., & Street, E. (2018). The dynamics of 'post-crisis' spatial planning: A comparative study of office conversion policies in England and The Netherlands. *Land Use Policy, 77*, 811–820. https://doi.org/10.1016/j.landusepol.2016.12.005

Ricketts, S., & Field, D. (2012). *Localism and planning.* Bloomsbury Professional.

RICS. (2016). *Future of policy and standards for low and zero carbon homes.* https://www.isurv.com/downloads/download/2068/future_of_policy_and_standards_for_low_and_zero_carbon_homes_rics#:~:text=Future%20of%20policy%20and%20standards%20for%20low%20and,and%20the%20recently%20withdrawn%202016%20zero%20carbon%20target

Sharman, L. (2016). Deprived communities missing out on Neighbourhood Plans says think tank. https://www.localgov.co.uk/Deprived-communities-missing-out-on-Neighbourhood-Plans-says-think-tank/41533

TCPA. (2018). *Planning 2020.* Retrieved from London: https://www.tcpa.org.uk/Handlers/Download.ashx?IDMF=30864427-d8dc-4b0b-88ed-c6e0f08c0edd

Wills, J. (2012). The geography of community and political organisation in London today. *Political Geography, 31*(2), 114–126. https://doi.org/10.1016/j.polgeo.2011.11.003

Wills, J. (2016a). Emerging geographies of English localism: The case of neighbourhood planning. *Political Geography, 53*, 43–53. https://doi.org/10.1016/j.polgeo.2016.02.001

Wills, J. (2016b). *Locating localism: Statecraft, citizenship and democracy.* Policy Press.

Wiltshire Council (2023). Climate Change Guidance Note for Neighbourhood Planning Groups. Climate Team, Wiltshire Council. Trowbridge.

Wootton, P. (2017). What does the presumption in favour of sustainable development mean? https://www.howespercival.com/articles/what-does-the-presumption-in-favour-of-sustainable-development-mean/

CHAPTER 4

Frome: Crafting 'Independent' Identities in the 'Home' of Localism?

The New Economy and Planning for Transitions: Some Key Things to Consider

Notably because—as we saw in the previous chapter—in preparing a Neighbourhood Development Plan (NDP), the contributions made by local actors, and who or what is enrolled or disenrolled in the process are telling of the extent to which *New Economy* actors can navigate local networks, and the reach of alterity in formal institutionalised processes of governance. Or as Davies and Featherstone put it, "how different trajectories of activity are combined and reworked" in arenas that bring together 'niche' and 'regime' actors (Davies & Featherstone, 2013, p. 244). That is, what effect do different groupings of actors' interests and identities have on reconfiguring the political landscape and its 'coding' (ibid.) to promote *placemaking transitions*? Understanding this enables us to explore how personal, place-based and organisation identities become embedded into the fabric of local government and alternative forms of local governance.

This book therefore aims to offer a new lens of the nexus between planning, placemaking, social movements and transitions. Set within a context of wider socio-technical transitions theory that seeks to explain the potential of 'scale shift' from the fringe, or 'niche' to influence mainstream, or 'regime' institutions, the case study of Frome will explore what happened in the town and why it happened. Starting with an outline of the town's

A. Burnett, *The Politics of Transition*, Palgrave Studies in Sub-National Governance,
https://doi.org/10.1007/978-3-031-40769-7_4

history and some of the key issues shaping development in the town, we introduce some of the main actors involved in shaping sustainable place-making and their role in shaping an "independent" identity for the town and how this shaped the wider demands for place-based political action by independent councils and the *Flatpack Democracy* movement. Our quest in using Frome as a case study is to uncover the conditions that led to a strong relocalised (circular economy) character in Frome's placemaking networks and the implications for Frome and the wider lessons for transitions to alternative development futures—how they can happen and their replicability.

In particular:

- What factors explain socio-political innovations in Frome?
- To what extent can Frome's experience be deemed as a 'sustainability transition' towards relocalisation?
- What are the implications for understanding the 'niche' and 'regime' in post-Brexit planning theory and practice?

Why Frome and What Is Its Significance to Community-Led 'Sustainable' Placemaking?

Frome lies within eastern Somerset, South West England—previously in the former Mendip district area and which is now covered by the unitary authority, Somerset Council (see Chap. 11 for a more detailed discussion on the implications of this change). The word Frome is derived from the Welsh ffraw, meaning running water, a reference to the River Frome on which the town stands. Nearby is the world-famous music festival location Glastonbury and one of the smallest cities in the UK, Wells. Frome has a long history as a market town and its economy traditionally centred around textile and wool manufacturing, later diversifying into metal work and printing[1]; its settlement dates back to 685 and its market is referred to in the Domesday book (1086) (MDC, 2008).

[1] Frome has a rich historic character with 373 listed buildings, the highest concentration in any town in Somerset and its unique spatial features include the cobbled street of Catherine Hill (Fig. 4.1) and a water course running through (medieval) Cheap Street—one of the central pedestrian corridors in the town centre. The town's historic value was recognised through the designation of a conservation area in 1973 and its distinctive features and their conservation was sought in the 2012 Town Design Statement (TDS) (see WYG, 2012).

Historically, Frome was the largest town in Somerset in 1801 (population 8500). The town experienced an upsurge in its population after the Second World War and in the 1970s (population c.20,000), partly as an effect of strategic planning to protect the green belt in nearby Bath and Bristol (MDC, 2008). The 2011 census calculated 26,203 residents—comprised of 11,198 households—making Frome the fourth largest town in Somerset with the highest population density in the county[2] (ONS, 2011; SCC, 2016). The 2021 census indicates that the population of the former Mendip district saw an overall rise of 6.2% since 2011 (though compared to other areas in South West this was a rather modest increase) while Frome's population increased to 28,559—over 2000 people—during the inter-census period (Though there seems to be some variation in these figures depending which source is used).

Frome has a relatively young population compared to other areas within the former Mendip district, comparatively attracting more people in their 30s (SCC, 2016), but the number of 15–29 year olds is well below the national average (ONS, 2021). As we will explore further, Frome has become a magnet for 'trendy' and alternative personalities seeking to live in a South West rural market town, also known locally as "New Frome" or "Refugees from the South East".[3]

The development challenges facing Frome are typical of many towns and cities across England, such as providing enough affordable houses and retaining and encouraging local businesses in the area. A study on Frome's employment sites showed that several sites had already been lost to housing, with a further 19 out of 25 sites at risk of future loss (Brunsdon & PJK, 2012). Frome's traditional manufacturing sector has been affected by "significant structural economic changes", leading to a loss of 1110 jobs between 1998 and 2004 (MDC, 2008). Frome has a train station

[2] Frome had a population density (per hectare) of 41.3, compared to 1.5 in Somerset county and 4.1 in England (SCC, 2016). In 2021, the UK census found that overall population density was 277.05 inhabitants per square kilometre but it was 157 in the former county of Mendip (see https://www.ons.gov.uk/peoplepopulationandcommunity/populationandmigration/populationestimates/bulletins/populationandhouseholdestimatesenglandandwales/census2021, last accessed 3.12.23).

[3] The recent decision to declare a housing emergency has been blamed on 'blow-ins' from London who have priced out local residents in a post-Covid housing market. (See for instance https://www.telegraph.co.uk/money/property/somerset-town-council-housing-crisis-blames-londoners/ and https://www.thetimes.co.uk/article/no-room-in-frome-are-locals-being-priced-out-by-londoners-cxfkffgjw, last accessed 3.12.23 - note that both websites require a subscription to access them).

which has a direct route to London Waterloo (as of 2015). However, evidence suggests that the town performs a dormitory role, where a significant number of residents live in Frome but work elsewhere.[4] Despite a strong 'green' identity increasingly associated with Frome, data compiled in 2016 found the use of the car as the main method of transport for commuters was high at 71.9%, compared to a 62.7% national average (SCC, 2016). At this time, the local economy as of 2016 was strong, with only 12 vacant shops out of a 188 (6.3% compared to the national average of 11.7%, SCC, 2016). The town had a low unemployment rate at 1% (SCC, 2016) and the highest concentration of jobs within Mendip, although its ratio of jobs to population is lower than other towns in the former district (MDC, 2008). As of 2023, the unemployment rate for Somerset stood at 2.6%.

Frome has an active network of community groups with a strong cultural attraction, including the annual Frome Festival, an independent cinema, two theatres and the international music venue The Cheese and Grain.[5] The town benefits from several significant natural assets including the open space adjacent to the centre of the town Rodden Meadow and the River Frome which cuts through the town and several parks and open spaces managed by FTC (Fig. 4.1).[6]

Frome's historic and physical environment has set the scene for a strong 'independent' local identity cultivated by several key actors. In 2010, John Harris (a journalist for the Guardian newspaper and a Frome resident) produced a short film "Will Neo-liberalism Eat My Town?" (Harris, 2010). Times of austerity and a battle over whether to have, or reject, a supermarket on a central and mainly redundant 4.76 ha site in the town (Saxonvale) made this seem a possibility (see Chap. 8 for more on the

[4] In 2011, 3714 more people commuted out of the town than commuted into the town for work (SCC, 2016), absorbed by neighbouring Warminster and Trowbridge in Wiltshire and the cities of Bath and Bristol, due to the close proximity to the A350 and the A36 (MDC, 2014, p. 14).

[5] A former grain warehouse in the centre of Frome which has recently hosted international bands, including the Foo Fighters. A Locality programme Power to Change briefing documented the refurbishment and the Cheese and Grain's success as a community enterprise https://www.powertochange.org.uk/wp-content/uploads/2018/03/The-Cheese-and-Grain-Case-Study.pdf (last accessed 11.1.19).

[6] Until initiatives by FTC and partnership with Sustrans, there were fragmented access issues to experiencing the river corridor. Some green spaces were under-utilised and subject to littering and anti-social behaviour. Frome Missing Links has also been campaigning to link the town's river corridor to the wider local footpath and cycle path network (see https://fromesmissinglinks.org.uk/, last accessed 3.12.23).

Fig. 4.1 Catherine Hill (Source: author)

recent Mayday Saxonvale campaign). Yet, in the words of one respondent Frome, like Totnes, is a place that is "quite unusual" in being a "community that has quite a strong sense of its own agency".

Frome has gone from being a relatively 'normal' market town to one that is bucking a recent trend of a decline in engagement in politics towards a more invigorated local democracy. Since 2011, the town council has been controlled by a group of independent councillors and there has been a consolidation of relocalisation and circular economy principles in the context of the broader localism agenda. Now, Frome is becoming increasingly renowned for its thriving independent identity, politically and economically. Frome itself has undergone a shift in its reputation as a slightly down-at-heel town (compared to other neighbouring Somerset towns), and relatively unknown outside of this area, to being defined by national media as one of the 'coolest' places to live. It is repeatedly mentioned in local and national newspapers as a trendy and trend-bucking resilient town. In 2016, Frome won the Great Town Award[7]—an accolade which mirrors its increasing standing and popularity at a national level. Frome was also named as the sixth coolest places to live by the Times newspaper in 2014.

The Sunday Times listed Frome as the best place to live in the South West and in 2018 and 2019 Frome was named as one of the best places to live in the UK—where its historic character now reflects 'trendy' credentials from a previously 'down-at-heel town' (see *The Sunday Times*, 2018). Independent boutique retailer Trouva.com named Frome as the most "stylish" place in the UK. In 2021, a Sunday Times journalist declared that "Frome in Somerset is the regional winner of its 'best places to live' award because of its a pretty cobbled town "that likes its shops the way it likes its politicians—independent" (Palmer, 2021). (Interestingly, they also suggested that an "independent spirit rules in the not-so-wild west" and recognised other Devon towns for their independent identities). Though, this accolade has continued to be repeated regularly, though interesting in 2022 it was knocked off the top spot—which may suggest a plateauing or waning interest in its emboldened reputation—though in 2023 Muddy Stilettos returned this title as the Somerset regional winner of the best place to live.

Frome boasts one of the largest independent markets in South West England and has seen a surge in the creative industries, such as marketing, design and small-scale start-ups, that have replaced (in part) the fall in traditional manufacturing jobs. A co-working enterprise hub was established in 2007 by an entrepreneur who moved to the town (latterly expanding to nearby towns and cities), which aims to be a hotdesking facility for "Independent and Creative Thinkers" in the centre of Frome[8] and was part of their vision to establish Frome as a "21st Century Market Town".

The town has also caught the attention of the national media elsewhere,[9] notably for its ambitious independent (and green) town council, which has been referred to as a 'woke town' (a term that has African American

[7] See https://www.frometowncouncil.gov.uk/frome-wins-national-great-town-award/ (last accessed 7.1.19).

[8] Forward Space and the former MDC have shown signs of exploring flexible work units across Frome. As of December 2023, there were five coworking spaces listed on the website www.coworker.com, indicating the expansion of co-working facilities in the town, reflected in many other UK towns (see https://www.coworker.com/search/united-kingdom/frome, last accessed 3.12.23).

[9] See https://www.discoverfrome.co.uk/frome/frome-in-the-press/ (last accessed 11.1.19).

Fig. 4.2 Frome Town Council Logo (2023)—highlighting how the council (and implicitly the town) is 'made differently'

origins referring to perceived awareness of social justice[10] that has more recently been captured by right-wing critics of socially liberal and inclusive approaches). This is noted in Frome Town Council's (FTC) recent logo iteration, with the multi-coloured strips, possibly to reflect recognition of the LGBT+ movement (Fig. 4.2).

No wonder the Frome Independent market's slogan is daring enough to claim: "In the Middle of Nowhere, Centre of Everything" to indicate the high levels of activity in the town in its predominantly rural setting. However, this popularity comes with a price—reflected in the sharp 14.5% annual rise in house prices between Dec 2015–Dec 2016[11] and during lockdown, prices rose by 13% between 2020 and 2021. (However, compared to the rest of Somerset county, Frome's property prices are relatively mid-range, compared to areas closer to the cities of Bath and Bristol, see PlumPlot, 2023). However, as a journalist from the Financial Times commented, there are concerns that the "tsunami from London", or the 'Down from London (DfLs)', is eroding local charm in rural towns and villages, noting that Frome is now "gentrifying rapidly"; arguably bringing both "a new life and wealth" to what were traditionally more rural communities, boosting the creative and local economic capital of different trades and some (social) enterprises but at the same time pricing 'locals' out of Frome's communities housing market (Dare Hall, 2021). (Refer to Chap. 9 for more discussion on the politics of gentrification in the town).

Below, we elucidate the relational and affective ties within Frome's placemaking networks, reflecting what Derman (2020) calls "the politics

[10] The term 'woke' is a relatively fashionable term to describe people and places that are attuned to issues such as social injustice and discrimination, that is, 'awoken'—implying a transcendental awakening to societal issues. See Britten (2018) for a visual depiction of the "Woke" town of Frome by the illustrator Jordan Andrew Carter. However, as we explore in the final chapter, such associations can also prove problematic for assumptions and prejudices around place and group identities.

[11] Compared to a rise of 8.5% in Mendip district, 6.0% in Somerset and 7.0% in the UK over the same period.

of connection". We assess the multi-dimensional relationships between these actors, the values underpinning groups involved in relocalised initiatives in the town and the role of informal ties, such as friendship, as a condition for shaping the governance of these networks—including how this shaped the democratic and political form of Frome's independent group of councillors and the initiatives they cultivated. Frome offers the opportunity to explore whether independent politics can transcend often conceived 'sticky' transition dynamics in planning practice. We also explore some of the issues relating to participation when a group of highly active people seek to re-design local institutions and the identity of a town to reflect deep green sustainable values.

Independents for Frome—Defining a Politics 'for Frome'

The wider constituency of Somerton and Frome was traditionally held by either Liberal Democrat representatives (notably David Heath from 1997 until 2015) or the Conservative MP David Warburton from 2015–2022 (Warburton resigned as an MP after he was charged with allegations of misconduct, triggering a by-election which was won by Sarah Dyke, and saw the seat return to the Lib Dems). However, beneath the scenes of a typical party-political dualistic battle between two main parties in this Parliamentary constituency lies a stirring of a new political movement in the town of Frome set within the context of an increasing vote share for independent and Green party candidates (more on this in Chap. 11).

Frome is pushing a sustainable place-shaping agenda in many innovative ways since FTC[12] became independently controlled when a non-party-political group—the Independents for Frome (IfF)—won a majority of seats in the council in 2011.[13] IfF now occupy all 17 seats in the council—having won a full house in 2015 which was repeated in the 2019 and 2022 local elections.[14] IfF is registered as a 'minor political party' by the UK electoral commission so it can include the name IfF on the ballot paper but vehemently shies away from being identified with 'party politics',

[12] For a summary of the responsibilities and objectives of FTC see http://www.frometowncouncil.gov.uk/council/frome-town-council/

[13] http://iffrome.org.uk

[14] A list of the current IfF councillors and their motivations for being a councillor can be found at http://iffrome.org.uk/meet-the-candidates. Motivations of IfF candidates in the 2011 election can be found in Macfadyen (2014). For a list of the councillors elected in the town council and district elections see https://www.frometowncouncil.gov.uk/council/frome-town-council/about-frome-town-council/elections/

which it sees as toxic and divisive (exemplified in a slogan used by FTC 'Progress without Politics' in an event in 2018 to reflect and deepen FTC's independent approach in 2018—though as we will see, even independent politics is intensively 'political').

IfF purposely rejects party politics as a "poison" that "infects" the capacity of political representatives to make fair decisions (Macfadyen, 2014). IfF's politics are claimed to be inclusive, emergent and draw upon maximising collective action where the town council plays a *facilitating* role as opposed to following the 'party line', which is the case in some town councils in England. IfF uses a set of formalised rules and values, the *Ways of Working* (WoW), to frame their decision-making which claims to respect the diverse personal views within the group and aims to put Frome's interests (rather than party politics) first. (Having won all 17 seats in the 2015 council, IfF sought to extend the WoWs to become the *modus operandus* for the council by including them within the standing orders to institutionalise these informal rules of engagement further). See Appendix C—IfF's Ways of Working.

There is evidence that IfF have contributed to an increase in political engagement, with 75% more people casting their votes in the 2011 elections than in 2007 (Macfadyen, 2014). In 2015, 50 candidates stood at the FTC elections.[15] One national level respondent I interviewed suggested that most independent councillors suffered losses in the 2015 election due people casting their local vote according to their affiliations to party-political candidates at higher tiers of representation on the ballot paper (i.e., local authorities and MPs). For many voters, the 2015 elections was three tier—national, that is, MPs, local authority and town council elections—the fact that Frome's gains in the town council bucked this trend was perceived as significant. (And as we will see, in the 2022 district elections, IfF not only won all 17 seats again but this time the Green Party candidates made significant gains, indicating the consolidation of local green politics at both town and higher tiers of government).

We home in around several significant events in Frome that actively contributed to the town's relocalised and independent identity of what has been branded as '*The People's Republic of Frome*' (Harris, 2015)[16] and we

[15] See https://www.frometowncouncil.gov.uk/council-elections-whos-who/ and http://www.frometimes.co.uk/2015/04/21/frome-town-council-election-guide/, which includes a summary of town council candidate motivations and goals within office in 200 words.

[16] It should be made clear that this 'launch' is considered here through the release of the newspaper article in The Guardian shortly after IfF won all 17 seats where local journalist John Harris declared 'How Flatpack Democracy beat the old parties in the People's Republic of Frome' (Harris, 2015).

journey into the effect political and environmental experimentation is having on local democracy and resilience within the town and beyond. Whilst this phrase 'The People's Republic' was coined by a Frome-based journalist writing in The Guardian newspaper, it depicts the deep sense of wilful autonomy that drove many Frome-based actors to engage in the government policy agenda of localism and to think creatively about how it could be applied, or subverted, to achieve radical and sustainable outcomes in the town of Frome and be a model for other towns to follow.

FTC had experienced a long-standing antagonism with the local authority, Mendip District Council (MDC)—a Conservative-led authority which has resulted in some value-driven conflict between the two levels of government and solidified an at times vehement anti-party-political stance amongst IfF councillors. As we will see, this antagonism between Frome-based actors and the Conservative-controlled district council was a key point of contention that sparked the motivation for some individuals to create and join IfF and to advance relocalised policies and initiatives through FTC.[17]

More recent changes to the administrative boundaries of representation are of key interest as the former MDC has been disbanded in favour of a unitary council, Somerset Council. This new council amalgamates the former district councils of Mendip, Sedgemoor, Somerset West and Taunton, and South Somerset, alongside the existing services provided by the former Somerset County Council (see in particular Chap. 11). The fact that there has now been a significant shift in relational ties among government administrations is of key interest in setting the context for how in each administrative arrangement local and district actors enacted claims to autonomy to define the conditions for development in Frome.

In the chapters that follow, we explore how these identities have mutated and been consolidated in subsequent IfF administrations and how salient challenges such as post-Brexit, post-pandemic recovery and unitary governance are affecting 'Frome's' ability to champion 'independent', sustainable and democratic renewal.

IfF and Its Discursive Ties to Localism and Transition

During a visit in February 2016 by Eric Pickles, the former Secretary of State for the Department of Communities and Local Government (DCLG), declared Frome the '*Home of Localism*' (Harris, 2015). A

[17] For an historical account of the vote share in Mendip district 2003–2019, see Teale (2007).

vigorous local narrative reflected within IfF's politics enabled FTC to latch onto to the policy discourse of localism—but with subversive objectives. Frome actors sought to go beyond the original intentions of localism to 'empower local people'. As Peter Macfadyen suggests in the publication *Flatpack Democracy* (Macfadyen, 2014), IfF latched onto potential for localism legislation to actively construct a new political model, drawing on the discourse of localism (a policy directive of central government) to orchestrate the scaling-up of parallel infrastructure in the everyday politics of the town. Frome's historical rooting of independence in the town's economic history was consolidated by the independent's group objective to erode party politics within the town council: "it has to be total revolution or nothing: that was our stance" (Macfadyen, in Harris, 2015).

Some of the town councillors who took power in 2011 from the perceived ineffective, incumbent town council were members of the local Transition Initiative, Sustainable Frome, including Peter Macfadyen as the "chief theorist" of Flatpack Democracy and Sustainable Frome founding member (Harris, 2015). Under IfF, FTC's approach is to fuse relocalisation and the government policy of localism, showcasing New Economy practices in the UK. The politics of IfF has been heavily influenced by the values and practices of *Transition* and radical international social organising in Spain's *Podemos* and other social movements that aspire towards Commons-based development and the New Economy. In effect, IfF originated as a political mutation of *Transition* which set the foundations of an emerging UK independent political movement for those that seek to emulate it. For instance, it embraces ideas such as parallel infrastructure, relocalisation (promoting positive 'sustainable' local resilience feedbacks and echoing circular economy principles), independence from political parties, as well as promoting shared assets and prefigurative politics (the enactment of the ideal in everyday practices). IfF has since ignited a broader movement of political change in the UK, several European countries, and the United States under the Flatpack Democracy "do it yourself" independent political banner. However, as we will see its own evolution as a movement has been hampered by the distribution of power at a local level, which Macfadyen suggests implies that local councils (i.e. town and parish councils) may not yield the potential for radical change he hoped it might (see Chap. 11).

The following table sets out a comparative analysis of Hopkins' assessment between localism and relocalisation with the additive feature of the political agenda of IfF to explore the key political entities (discourses and governance) involved (Table 4.1).

Table 4.1 Comparative analysis of Hopkins' assessment between localism and relocalisation and IfF's independent political agenda

Aspect	*Localism*	*Relocalisation*	*IfF*
Governance	Formal procedural mechanisms, e.g., LPA and Examiner, referendums, planning applications, asset registration. Invites consultants as advisors. Reducing central government influence in local matters.	Follow Transition Networks's steps to transition (see Transition Network, 2018). Stronger and more accountable communities and government—all tiers.	Deliberative governance (see Macfadyen 2014). Governance with the community, local government an enabler of community-led initiatives. Stronger and more accountable local government.
Economy	No decoupling from the globalised economy. No shift in role of consumer and producer.	Decoupling from globalised economy. Local infrastructure tied to local provision and consumption.	Local infrastructure tied to local provision and consumption. Support for relocalisation and local (socio-economic) sustainability.
Development	Industry-dominated; Some policies in favour of self-build; increasing interest in cohousing, net-zero.	The Commons (shared assets); relocalisation; cohousing, self-build; sustainable design.	The Commons (shared assets); relocalisation; cohousing, self-build, sustainable design.
Community	Importance of community ownership—transfer of assets from state to community.	Importance of community control and ownership. Communities take action to develop activities beyond the state.	Importance of community ownership. Communities take action to develop activities beyond the state and with the (local) state as an enabler.
Power/ political	Local people should have more control over decision-making and services. Lack of feedback mechanisms.	Local people should have more control over decision-making and services. Apolitical. Parallel infrastructure beyond the state. Emerging endorsement of Transitioners into independent politics.	Radical, parallel infrastructure within and beyond the state.

(*continued*)

Table 4.1 (continued)

Aspect	*Localism*	*Relocalisation*	*IfF*
Climate Change	Lack of robust policy framework—though climate emergency declarations, helping to corral local climate responses. Lack of strong ethics on sustainability.	Strong ethical and organisational identity tied to addressing climate change in positive and intentional ways.	Strong ethical and organisational identity tied to addressing climate change in positive and intentional ways. One Planet Living; Early mover in net-zero target
Networks	Local levels of governance.	Part of *Transition* and New Economy movement.	Indirect strong links to *Transition* through some of its members; organisation within the New Economy movement.

Source: Adapted from Hopkins (2010), with additional content on IfF by the author

Recent FTC Achievements for Sustainable Placemaking in Frome

FTC has established many bold and ambitious initiatives since 2011—both inside and outside the remit of urban planning. Under IfF, FTC set out to institutionalise some of the abovementioned ideas not only as instruments of the state, but also as an agenda with which to foster partnerships with broader emerging initiatives to promote the Commons (principals of shared resource management) and relocalisation. Recently, key actors have sought to repurpose and establish Frome's identity, such as a '21st Century market town', to prefigure (actualise) a relocalised economy and lifestyles of those in Frome, emphasising its creative capital and potential to support enterprise through co-working spaces.

FTC's areas of focus (objectives, see below FTC, 2023) highlight the scope of its intent:

- Building a sustainable economy for the town.
- Building a vibrant community that is able to participate in local community life and decision-making about public services.
- Creating a thriving town centre.
- The delivery of efficient and effective public services.
- An innovative and supportive Town Council.

FTC has supported several events encouraging relocalisation and local engagement in the polity such as Democracy Day, Independence Day and One Planet Sunday—shared objectives which are central to IfF's (non-) party-political and green agenda. In 2016, FTC even made the bold announcement that the town would become fossil fuel free by 2046, which was superseded by the target to become net-zero by 2030, following a broader campaign for decarbonisation in 2018 in the wake of the Paris Agreement and the later movement to declare climate and ecological emergencies.[18]

FTC's approach has been to crowdsource local knowledge and resources to boost socio-environmental regeneration from a commons approach, community-led stewardship of resources for the 'good of all'. For instance, in 2014 it supported a crowdfunding initiative to save a local field from development and co-facilitated the first Share shop in the UK[19] (where people borrow otherwise unused goods for a small fee, established in 2015), and from 2016 a local solidarity 'Community Fridge' that has the support of local businesses to promote a circular economy, which has the support of local businesses to promote a circular economy to prevent food waste. Both the Share Shop and the Community Fridge were initiatives of Edventure, which aims to be a "school for community enterprise" and provides courses to support young people in positive livelihoods and support community by creating enterprises that promote a sustainable economy—both Edventure and IfF have close ties to the New Economy. FTC sought to work with Sustainable Frome to create a by-law to endow the River Frome, which cuts across the centre of the town, with nature rights[20]—though unfortunately, this innovative approach didn't get the necessary support from government bodies to gain traction at that time.

[18] See https://www.frometowncouncil.gov.uk/your-community/resilience/clean-future/

[19] https://sharefrome.org/

[20] In 2016, Sustainable Frome approached FTC seeking to replicate innovation elsewhere in the world and responding to concerns over Brexit's potential effect on environmental resources; the group wanted to see if it was possible to grant the River Frome with legal rights conferred to it in the same way as an individual or registered organisation (known as 'nature rights'). FTC has worked with the organisation Nature Rights to draw up a draft by-law which was submitted to the former Ministry of Housing Communities and Local Government for consideration (see FTC, 2016) http://www.natures-rights.org/ (last accessed 6.6.19). For more information on creating a by-law see https://www.gov.uk/guidance/local-government-legislation-byelaws.

Frome has also sought to use the planning system creatively for positive development, through its Neighbourhood Plan (made October 2016) and exploring other rights made 'available' under localism including a Neighbourhood Development Order (NDO) and a Community Right to Build Order (CRtBO). Frome's Neighbourhood Plan was initiated in 2012 as a response to a call for Frontrunners under the Government's localism agenda[21] to be one of the first communities to develop a plan. Some key innovations and experiments included FTC's use of the neighbourhood planning process to promote strongly relocalised ideas and practices on cohousing, self-build, shared space schemes and One Planet Living (OPL) principles.

Under IfF, FTC also has a Youth Mayor and a Youth Deputy Mayor who seek to engage younger people in the town's politics and placemaking initiatives. IfF has increasingly sought community participation within its own governance.[22] For instance, a new online platform was launched to complement FTC's participatory budgeting initiative—known locally as 'The People's Budget'—that allows the community to suggest their own community project and to vote on other already suggested ideas. Initiatives such as these show the potential for radical localism in action—yet it is essential to understand who was involved in getting such initiatives off the ground, its wider relational and historical context and whether other towns and communities can realistically achieve the holistic offer of equitable and sustainable action by a pioneering town council—and indeed the extent to which the participatory and inclusive offer purported by Frome-based actors is indeed what it says on the tin.

The salience of some key events to the study of Frome's relocalised placemaking transitions is explained in Table 4.2.

[21] Frome was one of 20 Frontrunners supported by the government to pioneer developing a neighbourhood plan.

[22] See various YouTube clips for an example of FTC meetings in action, e.g. where Councillors sit 'cafe style' with the public in a in full Council meeting in December 2016 (see the FTC YouTube channel: https://www.youtube.com/channel/UCXUizztGFe9zjenOqxe_ABg/videos (last accessed 11.6.23), which also shows the town council's experimentation with participatory budgeting.

Table 4.2 Key events in the study of Frome's relocalised and independent placemaking

Key event	*Salience to the study of Frome's relocalised and independent placemaking*
2006—Establishment of Sustainable Frome	Sustainable Frome was established by Peter Macfadyen, in parallel to the *Transition* movement. Its simultaneous separation and affiliation to the wider movement indicates the strength in local vs. social movement belonging.
2011—The election of IfF to FTC	Winning 10 out of 17 seats, IfF's success can reveal the conditions to win power and how incumbents are squeezed out by the incomers within the structures of the state.
2013—The establishment of the Neighbourhood Planning steering group	Indicates how a set of actors coalesce and interact when party-political and independent councillors engage in planning arenas.
2015—IfF's win of all 17 seats at FTC	Can help to reveal the conditions that enabled IfF to consolidate power as protagonists.
Sub-event—Greens win 3 seats at MDC in 2015 election (but not stand at local level)	Indicates the broader assemblage of fringe political values influencing mainstream debates and channels of representation around the time of the 2015 election.
Sub-event—UKIP win 12.6% of national vote share[a]	
2015—The coining of the phrase "The People's Republic of Frome"	A phrase used by journalist John Harris (Frome-based writer for The Guardian newspaper) shortly after IfF won all 17 seats that indicates the town's perceived degrees of political autonomy in the wake of IfF's first full-house win of town council seats.
2016—The adoption of Frome's NDP	Can shed light on the political contestation between actors in planning arenas seeking alternative development models.
2016—House prices increase by 14.5% in a year (Dec 15–Dec 16)	Highlights the tensions of a changing socio-economic profile attracted to Frome as it increases in popularity and eco-cultural identity ties in the town.
2019—Launch of Mayday Saxonvale campaign	Owners of the Silk Mill (a creative workspace venue on the Saxonvale site) launched an initiative the 'Mayday Saxonvale' scheme to challenge a proposed 'business-as-usual' planning proposal being put forward by the developers Acorn Property Group (the latter was supported by the former district council and now unitary council).

2022 (May) IfF win another full house in the local elections, with only one out of eight wards contested	IfF councillors were elected without their seats being contested in all but Berkley Down Ward. In the seven other wards, only IfF put forward the necessary two candidates required for their ward—meaning these candidates were elected uncontested as there was no other competition. See https://www.frometowncouncil.gov.uk/council/frome-town-council/about-frome-towncouncil/elections/ This is extremely interesting as a case of 'democratic' representation—where a non-political party that emphasises participation and inclusion has reached such a popular standing in the town that no other political parties or candidates outside of IfF bother to stand. In Berkley Down Ward, a former Conservative town councillor tried to stand as an 'independent'—reflecting how IfF's rejection of party politics has filtrated into the political identities of other (traditionally party—political) candidates, such that there has been an almost total rejection of party politics at the town council since IfF came to power in 2011 (We discuss this more in Chap. 11).
2022 (May) Election of four Green Party councillors and the standing of two IfF candidates at the unitary elections for the new Somerset Council	In the new unitary Somerset Council elections, across the wards of Frome East, Frome North, Frome West the Green Party saw the election of four councillors. Some Green Party councillors also had roots in the transition (town) initiative, Sustainable Frome. Interestingly, two IfF candidates stood for election to the new unitary authority, indicating its experimentation in scaling up its independent political model from town to unitary council levels. In Frome East, the IfF councillor came in joint third place, narrowly missing election (which was taken by two Green Party candidates). See https://www.somersetlive.co.uk/news/somerset-news/localelections-2022-south-somerset-7046261

(*continued*)

Table 4.2 (continued)

Key event	*Salience to the study of Frome's relocalised and independent placemaking*
2022 (December)—Approval of the Mayday Saxonvale scheme with full planning	Former Mendip District planners granted full planning permission to the Mayday scheme.[b] Includes the UK's first Tiny Homes community and community owned assets on the scheme. "The UK's First Community Led Masterplan".
2022 The Frome Neighbourhood Network was established	The Frome Neighbourhood Network (FNN) aims to create a 'neighbourocracy' centred around empowered, resilient, caring and sharing communities networked on a street-by-street basis. The FNN was inspired by the strong community-/street-based support networks that sprung up during the COVID-19 pandemic. Annabel Macfadyen, Peter's wife, was a co-creator of the network and collaborated with FTC to roll out the scheme. Annabel sees the FNN and its neighbourocracy model as part of wider plans for a Neighbourhood Democracy Movement (which stemmed from India and supports sociocratic decision-making principles), which the group sees as a way organise local democratic mutual aid. If successful, this neighbourhood network could pose many possible connections to enhance and scale out meaningful participation in community-led planning (e.g., Neighbourhood Plans) and placemaking beyond Frome.
2023 (April) Abolition of Mendip District Council (MDC) and the creation of the unitary council, Somerset Council	On 1 April 2023, MDC was abolished and replaced by Somerset Council, a new unitary authority for the area previously served by Somerset County Council. This brought together the services previously provided by the four district councils in Somerset (Mendip, Sedgemoor, Somerset West and Taunton, and South Somerset). As such, previous administrative ties between the town and the district council were severed though quite a few individuals were absorbed into the unitary council (including planners).

2023 (September) Announcement that there will be a review of Frome's Neighbourhood Plan almost ten years since it was adopted	While there is no requirement under the NPPF to review and update a neighbourhood plan—unlike every five years for local plans—neighbourhood planning groups are generally encouraged to ensure their plan reflects current policy and to check it remains fit for purpose (reflected in planning practice guidance, Paragraph: 084 Reference ID: 41-084-20190509). Frome's NDP was approved 'made' in 2016. Since then, there have been numerous policy and administrative changes (including a move towards unitary governance in Somerset and the abolition of MDC). Though it was seen as 'groundbreaking' at the time, various elements need to be updated (including housing, business, employment, river corridor, public open space, transport and design). The original 2016 plan also had no standalone section on sustainability (for reasons that will be explained) and it doesn't refer to the climate and ecological emergency, which emerged into climate action discourse more recently. There are also calls for further strengthening of the 'sustainable development' underpinnings of a forthcoming revised neighbourhood plan, including items that were removed from the original drafts of the neighbourhood plan, such as tree planting, or focusing on the 'softer' side of planning policy, such as well-being issues.
2023 (October)—the Mayday Saxonvale alternative scheme was upheld at judicial review	The Mayday Saxonvale campaign team took the disputed Acorn Property Group scheme for Saxonvale to court under a judicial review process. The judge ruled in favour of the Mayday Saxonvale's case—disputing the lack of employment provision amongst other things—and officially quashed Acorn's scheme. The Mayday Saxonvale website includes one of its members explaining the effect of judicial review on the scheme. It may be that Acorn can rework and resubmit their proposal to Somerset Council (for a third attempt at getting approval for their scheme), which remains to be seen. In the meantime, an outline application for the Mayday Scheme has been approved and the campaign team will be working with planning and design professionals to build the UK's first community-led masterplan that can deliver a creative scheme that also includes community investment models to support wider investment in Frome's economy and local skills development (see Chap. 8).

[a]The principal event and sub-events are analysed simultaneously since they represent broader political phenomenon at the 2015 election (local, district and national). A concern of IfF councillors had been that voters would align their vote with other political party candidates at national and district elections. This clearly was not the case with IfF winning all 17 FTC seats in 2015 (and 2019)

[b]See https://maydaysaxonvale.co.uk/

Conclusion

In this chapter, we have begun to explore how elements of relocalised and political innovations can shed light on the influence of place identity and personal and organisational relationships as they intersect with broader phenomena (i.e. policy agendas at different scales, wider social movement ambitions and changing socio-economic conditions).

One respondent in interviews carried out for this research suggested that "Frome is just starting to make its mark and from the outside it's working"—alluding to more political complexities behind what might be presented; while another interviewee spoke of the intangibleness of the ever-increasing embeddedness of relocalised action in the town, meaning it has become less 'tangible' and able to pin down how and where such action takes place and the relationship between different entities:

> *It's, it's almost, it's, it's got to the stage where [long pause] rather than being a sort of concentrated minority of things, it's exploded and it's everywhere and it's therefore less tangible, somehow.*

In the chapters that follow, we seek to locate the conditions underpinning agency and resourcing of innovations in Frome within and outside the planning process and reflect on what this means in the context of broader changes to the administrative foundations between tiers of government, the cost-of-living crisis and the risks of gentrification (more on this in Chap. 9 in particular). To seek to unravel these complexities to elucidate light on the qualities of niche-regime interactions to advance more radical and alternative placemaking, it is essential to understand the formal and informal relationships, contributions, strategies, and experiences of those engaged in broader placemaking activities, which we turn to next.

References

Britten, F. (2018). Welcome to Wokeville: How Frome, Somerset, became the unlikely epicentre for the conscious hipster. https://www.thetimes.co.uk/article/welcome-to-wokeville-how-frome-somerset-became-the-unlikely-epicentre-for-the-concious-hipster-vfl3tl5s8

Brunsdon, M., & PJK. (2012, November). Employment land study. Available from: https://www.frometowncouncil.gov.uk/council/frome-town-council/

planning-for-frome/fromes-neighbourhood-plan/neighbourhood-plan-process-documents/

Dare Hall, Z. (2021, May 25). The market goes 'ballistic' as the Notting Hill set decamps to Somerset. *Financial Times*. https://www.ft.com/content/a6cefb6d-8c21-42f9-bd84-cf4fbd4190dd

Davies, A. D., & Featherstone, D. (2013). Networking resistances: The contested spatialities if transnational social movement organizing. In W. Nicholls, B. Miller, & J. Beaumont (Eds.), *Spaces of contention: Spatialities and social movements*. Ashgate.

Derman, B. B. (2020). Mapping the politics of connection. In B. B. Derman (Ed.), *Struggles for climate justice: Uneven geographies and the politics of connection* (pp. 183–219). Springer International Publishing.

FTC. (2016). A neighbourhood plan for Frome 2008–2028. *Frome: Frome Town Council*. Retrieved from https://www.frometowncouncil.gov.uk/council/frome-town-council/planning-for-frome/fromes-neighbourhood-plan/neighbourhood-plan-process-documents/

FTC. (2023). About Frome Town Council—Areas of focus. https://www.frometowncouncil.gov.uk/council/frome-town-council/about-frome-town-council/

Harris, J. (2010). John Harris in Frome: Will neoliberalism eat my town? *You Tube: The Guardian*. https://www.youtube.com/watch?v=VozcSgDiS0s

Harris, J. (2015). How Flatpack Democracy beat the old parties in the People's Republic of Frome. *The Guardian*. https://www.theguardian.com/politics/2015/may/22/flatpack-democracy-peoples-republic-of-frome

Hopkins, R. (2010). *Localisation and resilience at the local level: The case of transition town Totnes* (PhD). University of Plymouth, Plymouth.

Macfadyen, P. (2014). *Flatpack Democracy: A DIY guide to creating independent politics*. Eco-Logic Books.

MDC. (2008). *A portrait of Frome. A background paper to help support public consultation on the preparation of the Mendip sustainable community strategy and local development framework core strategy*. Retrieved from Shepton Mallet.

MDC. (2014). *Mendip Local Plan Part I*. Mendip District Council.

ONS. (2011). *Frome census data from HMG*. Office of National Statistics.

ONS. (2021). Build a custom area profile—Frome. https://www.ons.gov.uk/visualisations/customprofiles/build/

Palmer, T. (2021). Revealed: The Sunday Times best places to live 2021. *The Sunday Times*. https://www.thetimes.co.uk/article/revealed-the-sunday-times-best-places-to-live-2021-2z6wx8q2d

PlumPlot. (2023). Somerset house prices map. https://www.plumplot.co.uk/Somerset-house-prices.html

SCC. (2016). *Frome Town Profile*. Somerset County Council.

Teale, A. (2007). Local election results 2007—Mendip. https://www.andrewteale.me.uk/leap/results/2007/318/

Transition Network. (2018). The Essential Guide to Doing Transition: Getting Transition started in your street, community, town or organisation. Retrieved from Totnes: https://transitionnetwork.org/wp-content/uploads/2018/08/The-Essential-Guide-to-Doing-Transition-English-V1.2.pdf

WYG. (2012). *Frome Town design statement*. WYG Environment Planning Transport Limited. Cardiff. Retrieved from https://www.frometowncouncil.gov.uk/council/frome-town-council/planning-for-frome/frome-town-design-statement/

CHAPTER 5

Transition and the Origins of Relocalised Planning in Frome

The Origins of a (Relocalised) Political Revolution—Key Initiatives in Frome

Even before the election of the Independents for Frome (IfF) councillors in 2011, the strong role of identity to Frome's relocalised agenda is expressed in an interview with former MP David Heath, The Guardian journalist John Harris suggested that: 'Free market capitalism tends to squash places like this... This place has actually kept its identity' (Harris, video, 2010).The visual depiction of the salient identities within Frome's relocalised, independent placemaking networks can help to reveal some of the entities explored within this study (which is further detailed in the table in Appendix A).

A. Burnett, *The Politics of Transition*, Palgrave Studies in Sub-National Governance,
https://doi.org/10.1007/978-3-031-40769-7_5

Key entities[1] have played a significant role in Frome's transition arenas and the contestation local approaches placemaking, including: Saxonvale (a central mainly derelict brownfield site in the centre of Frome), Sustainable Frome (the local Transition Initiative, being the most notable organisation applying relocalised placemaking principles), IfF (as an independent political group with a strong sustainable orientation) and Frome's Community Plan ('Vision4Frome' V4F, which pre-dated the Neighbourhood Plan, the Neighbourhood Development Plan, NDP) as well as the Neighbourhood Plan itself. Many of these key entities can be classified as having 'roots in transition', that is, where more than one member of Sustainable Frome was involved in its establishment and/or implementation.

[1] That is, organisations, individuals and material phenomena which link phenomena that relate as part of a greater whole to Frome's relocalised placemaking (as per Edwards et al., 2014).

Mapping the relations between these groups is important to understand how they link formally—and informally—to produce local initiatives and shape the identity of placemaking approaches in the town. Because this is a study on the interaction of relocalised groups on formal institutional processes—that is, how they work with, through and around them to achieve their goals—relational ties and key local artefacts (such as community-led planning instruments) is of key importance. Some of these entities had a direct or indirect influence on Frome's NDP, or its predecessor the Frome Community Plan. Below, these relationships are further elucidated to indicate the changing relationships between (organisational and physical) entities from the period 2006–2023 and the nature of their relationships as being:

(i) independent—both entities can exist without the other
(ii) dependent—one or entity needs the other to exist
(iii) broken link—where a relationship was in conflict
(iv) no link indicates no explicit relationship at that time.

The date at which they were enrolled into Frome's relocalised network is also indicated.

Crucially, for this study the cofounder of IfF, Peter Macfadyen and author of Flatpack Democracy (Macfadyen, 2014), established Sustainable Frome in 2006 (in tandem with the first *Transition* initiative in Totnes in 2006) which acted as a springboard for relocalised initiatives in the town. Sustainable Frome became part of Transition Network in 2008, broadly following the necessary steps to *Transition* that new initiatives were meant to (broadly) follow. At this point in time, relocalised initiatives were not formalised until Peter Macfadyen started Sustainable Frome, which later became part of the *Transition* movement in 2008. There were no links between Frome Town Council (FTC) and Sustainable Frome at this stage (Fig. 5.1).

Sustainable Frome has met monthly since its inception and about 30–40 members regularly attend these meetings, although *active* membership dropped off when many key actors became more actively involved in IfF from 2011 and latterly when some members left the group to become part of Extinction Rebellion (XR) after 2018. There are approximately c.400 'members'[2] on its newsletter distribution list. For its organisation objectives (known as 'The Objects'), see Appendix B.

[2] 'Member' is used in a loose way to describe someone who attended Sustainable Frome meetings; as a network engagement was more on an informal basis and purposely lacked formal associations with 'membership' roles.

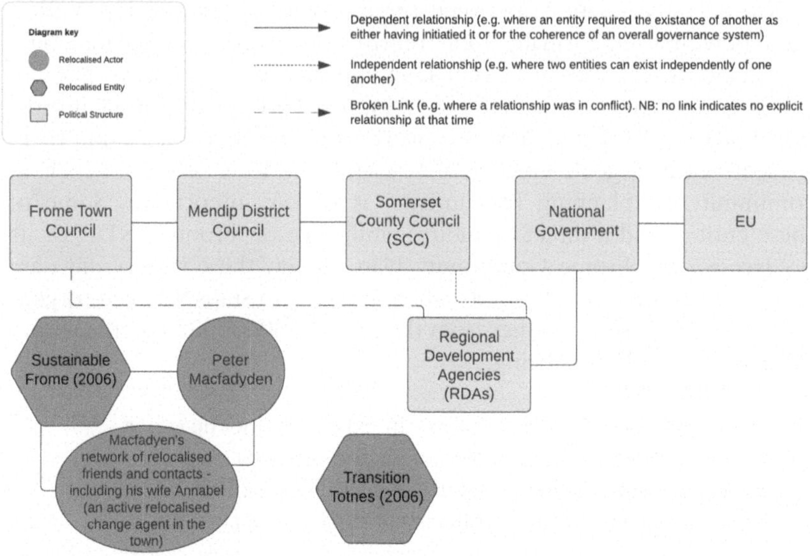

Fig. 5.1 2006—The establishment of Sustainable Frome

Sustainable Frome had extensive influence on the Community Plan and influenced the initiatives included as action points to carry forward. Its members participated in focus groups as well as the coordinating committee. The Community Plan also featured within Sustainable Frome's broader Energy Descent Action Plan as one of many strategies to reduce carbon emissions in the town. The Community Plan was later to become a de facto manifesto for IfF upon their election in 2011 and it became a key entity in shaping Frome's Neighbourhood Plan, discussed in the next section.

Despite the non-partisan stance of the movement, some members who had ties to local government helped to act as an interlocutor between grassroots environmentalism and government—partly through the creation of the Community Plan.

Sustainable Frome members also played a key coordinating role between different levels of government, including the local Member of Parliament (MP), Somerset County Council (SCC), Mendip District Council (MDC, the former district level local authority serving the Mendip district of Somerset) and Frome Town Council as well as with the broader

Transition movement. MDC had previously been supportive of several placemaking initiatives in the 1990s–2000s, including promoting town centre regeneration.[3] The former MDC Regeneration Manager Katy Duke (who lived just outside of Frome) was also involved in supporting various resilience initiatives in the town through her multiple roles in Sustainable Frome (as a former Sustainable Frome Director) and MDC.[4] Notably, she enrolled Sustainable Frome (amongst other community groups) to become involved in Frome's Community Plan.[5] However, during IfF's first administration there was a souring of relationships and a number of political tensions between FTC and MDC, which are explored in more detail in Chap. 6.

By 2010, when IfF was established, Sustainable Frome had formally become part of the Transition Network and a number of sub-groups existed, including the sub-entity the Greener Buildings group—a member of whom was the architect for the refurbishment of the Cheese and Grain. Some embryonic groups had already begun to splinter off from Sustainable Frome, including Frome Cohousing (in 2009) and Keep Frome Local (2012) (see Fig. 5.2). Frome Cohousing aims to promote community living in Frome, identify and procure suitable sites for community housing projects for Frome residents, and support the process of defining and formulating the management and policy structures any individual projects.

Keep Frome Local was a campaign group that had many Sustainable Frome members involved in an effort to determine how a mainly derelict site in the middle of Frome 'Saxonvale' (4.76 ha) should develop and what constitutes the 'right' kind of development in the town. This discursive battle was fought between two camps: Keep Frome Local (which campaigned for locally-led development that protects independent, local traders, i.e., relocalised) and Frome for All (a pro-supermarket entity, i.e., commercial and 'mainstream'). Keep Frome Local used a MDC-commissioned Saxonvale Masterplan (published in 2005) to justify their

[3] See https://fromeshops.wordpress.com/2015/07/27/regeneration-of-frome/ (last accessed 24.1.19).

[4] Indeed, pre-IfF and in the early days of the *Transition* movement a commitment by SCC sought to be the first Transition Local Authority, aims which subsequently were weakened with a change of political leadership (see Mitchell, 2011).

[5] A funding application from the Market and Coastal Towns Initiative (MCTI) was pursued after an unsuccessful application to the South West Regional Development Agency.

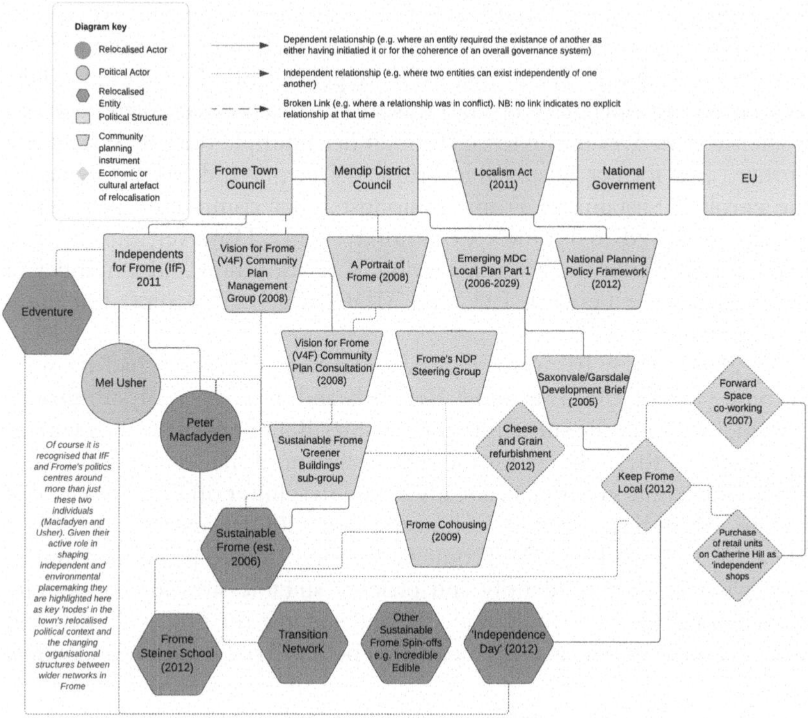

Fig. 5.2 2012—Following IfF's election in 2011

campaign, indicating how it used the structures of 'regime' planning instruments to support relocalised economic objectives in the town.

A more recent development on determining the right kind of development is reflected in the 'Mayday Saxonvale' scheme, which also involved individuals that had previously been members of Sustainable Frome. We will discuss more about the politics of securing relocalised planning alternatives on Saxonvale and in Chaps. 7 and 8.[6]

IfF itself was also arguably a Sustainable Frome spin off group where three of its members became town councillors in 2011 (see Table 5.1). Many spin-off groups listed below have invoked (governance) technologies of *Transition* and the New Economy. Recent key issues discussed by

[6]Local photographer Tim Gander has captured the changing face of Saxonvale of the online portfolio entitled "What Happened Here", available at https://www.timgander.co.uk/portfolio/saxonvale-what-happened-here/ (last accessed 8.6.19).

Table 5.1 Previous experience of IfF councillors (both 2011 and 2015 administrations) as either having links to Sustainable Frome or experience of working in government either as a councillor or in a professional capacity

Election	*Male*	*Female*	*TOTAL*
2011—elected (previous or existing members of Sustainable Frome)	2	1	3
2011—elected (former government experience—as either a councillor or as staff)	4	1	5
2015—elected (previous or existing members of Sustainable Frome)	1	2	3
2015—elected (former government experience—as either a councillor or as staff)	3	2	5

the group and related initiatives include resilient post-COVID recovery, affordable housing and renewed vigour on development approaches for Saxonvale under the Mayday Saxonvale campaign.

Some IfF councillors and FTC staff were or remain members of Sustainable Frome (see below) or had (significant) experience of working within different tiers of government. Table 5.1 indicates the number of IfF councillors with links to *Transition* and their previous role occupation within local government (either parish, town, district or county level):

Another Sustainable Frome spin-off group that demonstrates the connection between sustainability, resilience and links between the town council is Frome Renewable Energy Co-op (FRECo), which was set up in 2014 to fund clean energy projects that benefit the local community. FTC were a founding member of FRECo—again—in part championed by the former leader of the council Peter Macfadyen and the FTC Resilience Manager as a Director.

The Community Plan had been approved in 2008 with over 3000 consultation responses, but the plan had not been formally adopted by the incumbent FTC. These were not used as a basis of IfF's electoral pledges as IfF were offering 'change' and a break with what had come before and had an emergent, open approach to local politics (as a niche entrant to the local political scene, a strategy that encouraged wider transformative change).[7] However, upon winning the majority of seats in the council (to their surprise) and without a 'formal' plan of what to do in power - having

[7] See IfF's Ways of Working in place of party-political organisational identities in Appendix C.

not occupied seats in a town council before - they used the proposals put forward in Frome's Community Plan (V4F) to set IfF's agenda once in power as a 'de facto' post-election manifesto, and these priorities were also used to shape the content of Frome's Neighbourhood Plan.

IfF's emergence coincided with the rise of relocalised economic enterprises in the town when an entrepreneur purchased many of the buildings on Catherine Hill (Fig. 4.1) as retail units. These are let out only to the most 'quirky' independent retailers to purposely contribute to the 'independent' economic character of Frome and the street has become a hub for independent trade. This entrepreneur also established the Frome Independent, staying true to Frome's historical roots as a market town.

Originally billed as the "Frome Super Market" (note, a play on what the term 'supermarket', in the more mainstream sense, means in Frome) and following the success of these pilots, the market received some grant funding from MDC and FTC. It was rebranded in 2014 as The Frome Independent. The market has now become one of the busiest markets in the South West and once a month between March - December the town centre's roads and carparks are closed off to host 10,000+ visitors, and now stretches across the whole town centre and the Cattle Market car park outside the Cheese and Grain on market days. During the COVID-19 pandemic the market operated online. The same individual who invested in the market was also responsible for establishing Forward Space, the co-working space for creative start-ups. Forward Space has also since branched out to other areas in the Bath and Somerset area. This particular local entrepreneur had a positive vision for redefining the role of market towns drawing on their historic roots to reverse patters of out-commuting and encourage local economic multiplier effects as, in the words of a local entrepreneur, to: "re-work the economic model of market towns [and] bring everything right back into the town centre."

By 2016 (Fig. 5.3), the complex relations between entities and the increasing initiative of relocalised actors, political innovators and the availability of community-led planning instruments indicates the extent to which IfF, relocalisation and localism have influenced the types of initiatives within the town. This suggests that the formal links of the relocalised ideas set out within Frome's Neighbourhood Plan (as detailed in the later section Frome's Neighbourhood Plan began to be institutionalised both as instruments of the state, but also as a framework to guide an agenda that fostered partnerships with a broader emerging initiatives that promoted the Commons and economic relocalisation in the town and with wider social movement ties.

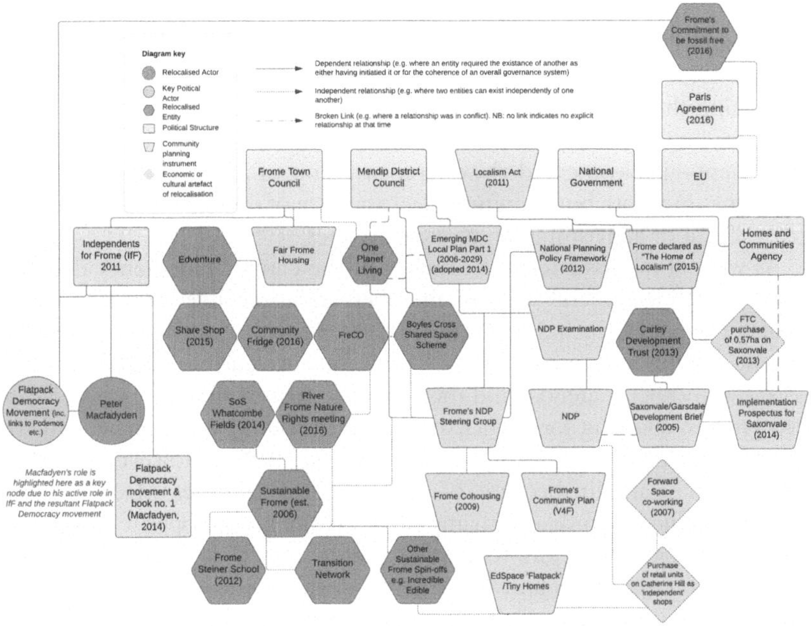

Fig. 5.3 2016—The adoption of Frome's made NDP

The global momentum for addressing climate change was reflected in the Paris Agreement, which FTC responded to through its commitment to become a fossil fuel free town by 2046.[8] These embedded of these networks led to a feeling in the town that Frome can deliver radical placemaking alternatives, leading one external New Economy actor I interviewed to suggest: "I love the way the stage is now set in Frome [to deliver on localism]."

Though established after the NDP had been adopted or 'made' in 2016, Frome's Extinction Rebellion (XR) group is important to understand how the ties to one social movement—Sustainable Frome (a Transition Initiative)—ebb and flow when new organisational identities emerge that reflect different qualities of environmental action (see Chaps. 10 and 11). In XR's case it offered a more radical form of (non-violent) direct action, which contrasts strongly with Transition's own

[8] As stated earlier, this later came to be superseded in December 2018 by a commitment to become fossil fuel free by 2030.

non-partisan approach which has kept it more of a bystander than a more vocal protest groups; such as XR who—in their earlier days at least—'took a stand' with a more visual, and controversial approach, blocking roads and gluing themselves to roads or even the House of Commons Speaker's chair (leading to various arrests). This has led some to charge the *Transition* movement—in Frome and elsewhere—as a 'bit of a talking shop', where new issues incubate but sometimes lacks a sense of urgency to take a stronger political or activist line on particular issues.

As we will discuss later, the urgency of the climate emergency debates around 2018 led several (former) members of Sustainable Frome to play a leading role in XR (including Peter Macfadyen's wife, Annabel Macfadyen), causing a hiatus in its membership. The two groups currently mutually support one another's agendas and there is clear cross-fertilsation between the two groups, allowing for participation in different types of 'niche' activist spaces in the town (see Dryzek, 2005; Melucci, 1989). However, it is also telling that (partially because of the influence of Peter Macfadyen in national XR governance and championing democratic as opposed to direct action), XR has also begun to assume a more 'softer' approach, favouring local climate assemblies and constructive engagement with local councils and communities to foster change—and rebellion—at a local level. There has also been a fusion of wider placemaking and well-being goals with the mode and technologies of transition; for instance the new Frome Neighbourhood Network (FNN) involves Annabel Macfadyen, and the group has adopted the *Transition* movement's 'Transition Streets' methodology to elicit a street-by-street approach to encouraging caring and resilient communities. More recently, wider community action is also framed within a new local government architecture of the new unitary authority, Somerset Council, and the broader declaration of climate and ecological emergencies which give greater fuel to Frome's relocalised placemaking apparatus. These recent changes in the political and environmental landscape in the town are discussed in the later chapters of this book.

Below, I detail the methodological approach taken to unravel the politics of the relational ties among actors and entities seeking to influence Frome's placemaking agendas highlighted in Figures 5.1–5.3).

Methodology

As a case study, Frome can help reveal important insights on the locally situated degrees of broader social movement ties—to Transition and the New Economy and the non-party-political 'indy' movement—within Frome's placemaking networks and how broader trajectories (notably localism) may influence these dynamics. For instance, how the town's placemaking dynamics weave into or are set apart from the structures of the state, are attentive to the concerns of the historical dynamics of place and how these inform the politics of local transition arenas, of which an NDP may be one of many. Meanwhile, exploring the duality of the 'niche' and the 'regime' within placemaking networks and both formalised and informal structures provide an extended view on the remit of planning on placemaking to solicit conditions for (relocalised) sustainable development.

While there are a number of other important actors and organisations in Frome, the main focus of this book is on the intersection of governance and practices between FTC, IfF, MDC, Sustainable Frome and the now unitary council (Somerset Council) to achieve relocalised placemaking transitions through politics and planning. This can be summarised within the framework of elements of the different 'tribes' of the New Economy (Table 5.2).

Table 5.2 Entities and phenomena under study within their respective 'tribe'

Element of the New Economy under study	*What are the key features of this element?*	*Principally examined through*
Transition and relocalisation	Initiatives for local self-reliance, appropriate culture, community resilience, drawing on the common bonds of place	Sustainable Frome, wider placemaking dynamics, IfF, economic relocalisation, Frome's spatial identity
Government and politics	Institutions and actors in the realm of public services, oversight, policy and regulatory frameworks	Local government structures and representation, party-political entities, the Neighbourhood Planning process, IfF, relational ties with MDC and other government entities (including the changing boundaries of administrative representation under the new unitary council, Somerset Council)

Source: Adapted from Real Economy Lab (2016)

Table 5.3 Respondents' roles and coverage in interviews

Role type/characteristic	*n*	%
All	27	100%
Male	18	67%
Female	9	33%
IfF councillor	4	15%
FTC staff member	4	15%
Neighbourhood Plan steering group	7	26%
LPA officer	2	7%
Neighbourhood Plan/FTC-commissioned planning consultant	4	15%
Member of 'Citizens Panels' (a focus group of local residents that fed into the consultation phase of the Neighbourhood Plan) who had either Sustainable Frome or Community Plan (V4F) links	6	22%
Regulation 14/ local consultee (in the preparation of Frome's Neighbourhood Plan)	6	22%
Former Sustainable Frome member	4	15%
Active Sustainable Frome member, e.g., as Director or attendee of monthly meetings	4	15%
Member of Sustainable Frome spin-off group (exc. IfF, covered above)	4	15%
Involved in a previous community-led planning instrument	3	15%
Actor external to Frome but part of relocalised and/or independent political networks	4	15%
Interview time total (hours)	32	N/A

In addition to significant desk-based research, a total 27 semi-structured interviews were carried out and transcribed. Those interviewed thus had a direct influence or involvement in the NDP process (directly or indirectly). The types of roles they performed in and around Frome's NDP are outlined in Table 5.3. This has very much been a longitudinal study, reflecting on Frome's experience over a ten year period (from 2014 to 2023), which can add to the richness of Frome as a case study.

Emails were sent out to those identified in a snowballing exercise, together with a research overview and summary. Many people interviewed as part of this research had been living in the town for over 20 years whilst others were relatively recent incomers, or locally referred to as 'New Frome'. Interviews' sought to gain an understanding of how interviewees became enrolled into Frome's relocalised identity network, including reasons for moving to Frome, or being contracted into the NDP (e.g., as a consultant), the contributions they made to the process and the challenges—and opportunities—they faced. All quotations or text that could reveal certain actors was first shared and discussed with respondents (in addition to measures to encourage anonymity).

A lot has progressed since the initial interviews have taken place (in 2014–2015) and as such various continued conversations with Frome-based actors were carried out to determine key results emerging out of Frome's Neighbourhood Plan, changes to the group dynamics within IfF and FTC, broader change pressures on the town, for example, from gentrification, land availability, inter-governmental relationships (especially with the new unitary Somerset Council), Brexit and the COVID-19 pandemic. In addition, I also engaged with those outside of Frome seeking to emulate the Flatpack Democracy model—successful or otherwise (see Flatpack Democracy—scaling up and out in Chap. 11).

The following actor types were sent a request for interview by email together with the research overview. Table 5.4 sets out the rationale for their involvement.

Table 5.4 Actor types interviewed and rationale

Actor type	*Rationale for interview*
All	• To understand the relationship between Sustainable Frome, broader placemaking networks, FTC, MDC and the planning process to date • Why Frome-based actors were attracted to work in or move to the town
All NDP steering group members	• To understand the role of personal values and the influences behind people's involvement • To understand the origins of innovative ideas and development alternatives • To understand steering group members' experience of the NDP process, governance and community engagement
Local authority representatives	• To understand Local Planning Authority (LPA) experience of the NDP process, governance and community engagement • To understand the difference between Frome and other NDP groups in Mendip district • Reflections on low-carbon policy context and relationship to the planning sector (e.g., developers)

(*continued*)

Table 5.4 (continued)

Actor type	*Rationale for interview*
Consultants involved in Frome's NDP	• To understand consultants' experience of the NDP process, governance and community engagement—including degrees of inclusion and exclusion and network ties (to *Transition* or 'regime' actors) • To understand the influence of consultants on the content of the plan vis-à-vis the policy context
Individuals identified as having links to Sustainable Frome *and* had sought involvement in the NDP *or* were engaged (e.g., Sustainable Frome members that also participated in the NDP Citizen Panels)	• To understand the role of Sustainable Frome in influencing the content and procedural aspects of Frome's NDP • To understand the relationship between *Transition*, FTC and MDC as well as broader social movement ties • To understand the origins of innovative ideas and development alternatives
External New Economy actors that had been identified as having links to Frome	• To understand the diffusion of any of the ideas set out in Frome's NDP regarding relocalisation and the New Economy to other places (or vice versa) to explore how pathways of transition were being carried, i.e., via the NDP process. • Probe partnership and network ties in the New Economy
National organisations representing core policy objectives in Frome's NDP	• To understand how Frome's experience sits within the national political context; notably how independent politics at different levels—town, district, national—is encouraging alternative models of politics • To seek alternative accounts for the sudden change in supportive policy context for community-led housing and self-build

With MDC now disbanded, I have been closely following the evolution of local democratic structures via various press releases, local media articles and secondary data relating to the changes since the new Somerset Council was created; noting in particular how this affects democratic functions between FTC and placemaking networks under these altered government administration configurations. In order to ensure that updates to this changing context were accurate, some of the key informants of this research provided useful commentary that has been used to reflect on changes arising since earlier drafts of this book.

Some national level actors were also identified and interviewed (e.g., individuals within the *Transition* movement or national-level independent political entities to complement a local perspective and offer insight into the institutionalisation of niche/regime aspects of socio-political transitions at different scales and the relative importance of Frome's achievements—particularly in relation to broader post-Brexit political change, administrative reforms and post-COVID-19 recovery.

The rest of this book examines in more detail how and whether localism is being used in a subversive, unintended way to up-end political, planning and placemaking norms in Frome and the potential for community-led influence on the politics of placemaking more generally.

References

Dryzek, J. S. (2005). *The politics of the Earth.* Oxford University Press.

Edwards, P. K., O'Mahoney, J., & Vincent, S. (2014). *Studying organizations using critical realism: A practical guide.* Oxford University Press.

Harris, J. (2010). John Harris in Frome: Will neoliberalism eat my town? You Tube: *The Guardian.* https://www.theguardian.com/commentisfree/video/2010/nov/30/john-harris-frome-neoliberalism

Macfadyen, P. (2014). Our 10 tips for reinvigorating local democracy—In one place. https://transitionnetwork.org/news-and-blog/our-10-tips-for-reinvigorating-local-democracy-in-one-place/

Melucci, A. (1989). *Nomads of the present: Social movements and individual needs in contemporary society.* Temple University Press.

Mitchell, E. (2011). Somerset Transition reversal raises questions over localism agenda. https://transitionnetwork.org/news/somerset-transition-reversal-raises-questions-over-localism-agenda/

Real Economy Lab. (2016). Real economy lab. http://reconomy.org/ [NB: be careful of malware on this page...]

CHAPTER 6

Transitions and the Niche–Regime Interface: The Politics of 'Independence'

The Politics of Representation—Conditions for Strong Placemaking Networks and Links Between Innovators of Relocalised Action

Frome's placemaking networks had strong degrees of multiple membership and often were the 'usual suspects' that were involved in getting (relocalised) initiatives off the ground. Representation in these networks was characterised by a limited group of people who referred to themselves as those with their "head above the parapet" who had strong affective ties by virtue of their love for Frome and were considered to expedite placemaking outcomes in the town. Many of the respondents enacting these roles indicated that they wore at least "two hats" owing to the multiple membership of different groups in the town and referred to themselves as the "usual suspects", a well-known term in planning parlance. Moreover, being involved in multiple local resilience initiatives was felt as a way of actualising interlocking green, social justice and civic engagement interests, reflected by the range of organisations in the town. This was seen as not a purposeful endeavour but borne out of being community-minded which helped to constitute these people as, in the words of one local group representative, "part of the group … we are

A. Burnett, *The Politics of Transition*, Palgrave Studies in Sub-National Governance,
https://doi.org/10.1007/978-3-031-40769-7_6

community-driven". Moreover, having a clear and strong view about issues of importance was also felt to increase social bonds between them.

The multiple memberships of these actors in the town and the intersection between informal and formal ties helped to create overlapping networks, which also helped to build the foundations of direct access to and shared understanding between placemaking networks and government. When the Independents for Frome (IfF—the non-party-political group of political protagonists seeking to shake up local politics) took control of Frome Town Council (FTC) as one respondent suggests:

> *A key factor in the way Frome has become more active in the areas you are interested in, and the areas that are strongly written into the Neighbourhood Plan, is because the group—a substantial number of the group [Sustainable Frome]—that were involved in environmental issues decided the town council were not taking Frome in the way that they wanted.* (local group representative)

As we will see, notably, pre-existing ties between some steering group members and the cohousing group led to greater levels of "receptivity" to include cohousing[1] in the plan. Pre-existing ties between FTC staff, IfF councillors and another group consulted in the plan also helped to create an "open connection" which led to mutual support and resources to fund a study that was used as part of the evidence base for the town's Neighbourhood Development Plan (NDP).

Some relocalised actors were particularly good at self-promotion, which amplified the identity of the town in this regard. The publicity of actions by key individuals in the local media and Frome's placemaking networks, or as a Sustainable Frome member explained, as "very good at appearing in the newspaper", was deemed a reason for elevating the prominence of some green initiatives.

Having skills, interests, time and capacity to be involved in placemaking and local politics draws from a "limited pool". Those involved in either Frome's placemaking or the Neighbourhood Plan were also deemed to be well-connected and able to leverage necessary resources

[1] Cohousing is a development alternative focused on shared living spaces by co-inhabitants (e.g. living room and gardens), in addition to personal, private spaces (e.g. bedrooms and bathrooms). For more, see https://cohousing.org.uk/.

with high levels of both social and human capital[2]—including time, skills and experience—for positive placemaking. However, the number of people who are prepared to participate in civic life to this extent was deemed to be from a "limited pool".

Friendship and kinship networks played a key role in forging relocalised networks and affective ties in Frome. When several Sustainable Frome members became IfF councillors, this was deemed to help forge close ties to FTC. One respondent described the bonds between actors as "incestuous" due to intersecting, multiple linkages—both between individuals and partners, who shared a "love" of Frome and a common vision of how they sought reinforced positive placemaking, as suggested by a Sustainable Frome spin-off group member. Some partners (couples) were found in a number of instances to share complementary professions, for example, a planner/builder, architect/architect, green pioneer/creative with one respondent describing one particular couple as a "powerhouse couple". For instance, when Macfadyen disenrolled from Sustainable Frome, his wife took over the role of coordinating it. *Two relocalised actors also lived next door to one another.*

Friendship was a key factor within Frome's relocalised placemaking network, with a Sustainable Frome member describing other individuals in the network as "extremely good friends" or "good friends"—friendships which cut across Sustainable Frome, FTC and other environmental groups. One respondent who later fact-checked my assumptions was keen to point out that whilst lasting friendships were built through Sustainable Frome, they questioned whether such enduring ties would follow with all fellow IfF councillors. Moreover, these friendships were built *through* Sustainable Frome, and did not precede it which necessarily influences both the tone and the trajectory of local placemaking ties.

One respondent described these ties as "multifaceted"—a combination of longer-term relationships, pre-dating professional involvement with FTC but also strengthened through a continuation of these relationships (friendships) in a formal context. Importantly, some individuals were friends *before* any professional involvement. Familiarity helped to strengthen the ties between placemaking groups and actors to navigate the state (the town council) when IfF occupied power as these community-government links were channelled through existing personal ties and shared visions of Frome's development trajectory. One respondent with

[2] See Putnam (1993) and Flora et al. (1996).

pre-existing ties (in virtue of their previous placemaking endeavours) talked of their relief when IfF came to power and shared frustration with formal processes to effect change. Although having direct access to FTC led to one local group representative feeling they had to keep a "low profile" as opposed to having their head above the parapet—indicating how visibility and influence are not necessarily the same depending on one's objectives and their positionality in local networks.

The Politics of Representation—Independents for Frome (IfF): Transitioning from Placemaking into a Flatpack Local Government

Cultivating the conditions for an effective route to power as a transition from Sustainable Frome (a local environmental group), the Community Plan (an instrument of community planning) and IfF (an entity to occupy the instruments of local government) meant the skills and experiences of those involved in a journey to take administrative control reconfigure the governance of the town council were critical. Below, we turn to assess the distinct '*conditions*' (positive and negative—highlighted in bold) that enabled IfF to take and consolidate power.

Path Dependence and the Instruments of Community Planning

Community planning instruments as a trajectory of sustainable placemaking (relocalised actors had skin in the game in Frome's Community Plan). The Community (parish) Plan (2008–2028), also known as Vision for Frome (V4F), was used by IfF once in power to set out their Strategy for Success 2011–2015. The Community Plan became a conduit for green issues that were corralled by Sustainable Frome throughout the process at various stages:

(i) Contributions from Sustainable Frome actively fed into the plan at the initial phases of the process when the plan was instigated by Katy Duke (former Mendip District Council (MDC) Regeneration Manager and Sustainable Frome Director).
(ii) When Macfadyen latterly became part of Frome's Community Plan Management Group and latterly as Chair.

One of the activating instances in the rise of IfF and its protagonist identity was because FTC (at that time) showed no interest in acknowledging (let alone implementing) the Frome Community Plan. The V4F Management Group had sought to inform FTC's Strategic Plan but the absence of formal links between them and FTC meant this diffusion of grassroots placemaking policy ideas did not occur. In 2010, when Macfadyen—who had also been involved in the V4F steering group as an interim Chair and as someone vehemently committed to sustainability issues—had asked the town council what their sustainability policy was, councillors replied that "the town had a park" and implied there was no need to look beyond this.

The personal resource (of Macfadyen) invested in the plan and the level of community support embodied in it—over 3000 consultation responses—was a necessary activator for political action and confrontation. This gave rise, in part, to dissatisfaction that led to the emergence of IfF. Having a community-wide agenda, rather than a party-political one, helped to foster a sense of an organisational identity that was focused on Frome (not party politics). Moreover, IfF's use of the Community Plan helped to lock-in Macfadyen's own relocalised networks into the formal structures of the state. The fact that the Community Plan had not been co-opted into formal governance, and it lacked an implementation mechanism, meant that IfF could use the Community Plan to shape its political agenda. The size of the town with a relatively large population (26,203 as of 2011, ONS, 2011), its well-established community networks and the disinterest of the incumbent town council in adopting or implementing the plan meant implementing capacity was not formalised, it was diffused across different groups in the town. A Memorandum of Understanding (MoU) between the management group and FTC was agreed for a period after the Community Plan was finalised, with the latter acting as a custodian of Community Plan. (However, the main custodian—the late Charles Ward—has since passed away and the Plan's caretaker governance has collapsed, partly because other community-led planning artifacts have since superseded it). The strength and activity of Sustainable Frome members to actively vocalise green issues lead to a skewed green tint which later set the tone for the conditions for the emergence of IfF.

There were already ties between community-based policies and the Local Plan, allowing for a fusion of grassroots-regime ideas and policies. As the Local Planning Authority (LPA), MDC was responsible for the local strategic planning policy framework (which is now performed by the unitary Somerset Council). As a parish plan, 'V4F' did not have a formal scope to influence planning decisions and it was not adopted as supplementary planning guidance. However, within the V4F process, there was evident cooperation and shared resource between MDC (as the LPA) and local actors as per their mutually reinforcing consultation agendas (see Fig. 5.2 in the previous chapter).[3] This collaborative exercise helped to lay the foundations of socio-political transitions. Firstly, it meant that there were strong, shared narratives in the Local Plan and the Community Plan; as one key person involved in the Community Plan put it, "so a lot of what is in the Community Plan is in the Local Plan as well".

Similarly, the regime-created entity of the Saxonvale masterplan (commissioned by MDC to set out design principles for the site, see MDC, 2005) was used as an artefact to enrol potential IfF councillors in the 2011 election campaign, given its resonance with design that was widely perceived as being appropriate for Frome. One IfF councillor described how they switched their preferences from pro-supermarket to anti-supermarket during an IfF-led event that used the brief to outline models for comprehensive development[4] of the site. Thus indicating how place-based identities can be shaken up by development alternatives and a clear vision and/or evidence base.

Role Boundaries and Degrees of Innovation

Lack of institutional experience at the town council level allowed for an open approach to campaigning. Prior to the 2011 elections, many ward seats were uncontested; that is many seats were 'co-opted' and not been contested by standing candidates—although interestingly the same thing has happened again under IfF in the 2022 local elections, as a result of their seemingly undisputed popularity and a 'ducking out' by other party political candidates (see Chap. 11). Under IfF there was an appetite to ignite a desire in some local people to stand as 'independents' and

[3] The Community Plan (V4F) was supported by the Coastal Market Towns initiative and MDC (with a £50k and £25k grant, respectively). Resources were used to support consultation for both the Mendip Local Plan and the Community Plan because the period coincided with the creation of the Local Development Framework, indicating the legitimisation of the local agenda by the district council. See Chap. 7 for more discussion on the governance of Frome's community-based planning instruments.

[4] Comprehensive development refers to the long-term planning for the different uses of a site and often may include a mix of housing, employment and amenity uses.

animate local politics. Interestingly, the political differences between the Conservative and LibDem councillors in 2011 meant they did not have a united front to challenge the new IfF-led town Council. In the words of one respondent, the other councillors from political parties "hated each other more than us" and tended to vote against each other due to a historically adversarial relationship that was defined by their party organisational identity (which tends to encourage oppositionality rather than cooperation).

The local structures of the state thus prised open to pressure from new (political) niche roles and practices. Such a break from the local political norm allowed for openness and agility in terms of what this new non-party-political group (IfF) could offer. An emergent approach was enabled because of a balance between a lack of the local 'regime' (i.e., at the town council level, since no IfF councillors had ever been a town councillor), but with sufficient insights from both social movement and regime institutions and roles (see below) to reconstitute how IfF could operationalise its ideas once in power: as one IfF councillor explained, "we didn't know what the issues are—we felt we couldn't say what we were going to do".

(Historical) tensions between Frome and the local authority allowed for the creational of oppositional identities to frame Frome's political and cultural difference. Party-political divides, cultural differences and historical tensions were felt to be acute points of contention between FTC-MDC relationships, not just between party political actors in the town council. Historical tensions and perceived cultural differences between FTC and MDC extenuated antagonism and role conflict, who were perceived to be disinterested and ignored[5] or were actively opposed to Frome's interests. One IfF councillor suggested that Frome's geographical setting "tucked away" in the north east of Mendip district exacerbated a sense of disconnection of the town from the district council (which was based in Shepton Mallet, in the middle of the former district boundary). A local group representative argued that the complexity of local politics was compounded by the three layers of local government and the party-political differences between them, as illustrated by this quote from an IfF councillor:

[5] For instance, according to one FTC representative, by not acknowledging comments on planning applications or changing the parameters of engagement midway through a process.

> *It's been really, really difficult, actually. Because I think they're so kind of culturally different from us, erm, I mean even other Tories in the region don't really like [short pause] the Mendip leadership [intake of breath] …. they're extremely aggressive, they're all men in grey suits—so they're not very much like us. Erm, they are [pause to thinks], they're really the opposite to us. It's so hard, I've never come across a more [short pause] difficult group of men to deal with, actually.* (IfF councillor)

IfF councillors felt a sense of duty as protagonists: to continually "*push the boundaries*" of the role of a town council and challenge its relationship to MDC. This sentiment was repeated by several IfF respondents, who considered themselves as persistent, challenging, creative, imaginative, more "progressive" and able to enrol expertise. MDC was often described in negative terms, considered to be one of many "dinosaur district councils" or "dysfunctional" and their councillors were deemed "nasty", "not nice people" and "aggressive" who were no longer equipped to respond to a changing community-driven planning context. Terms such as "God help us" were used at the mention of the district council and its duties to approve planning applications for Frome. A Sustainable Frome spin-off group member referred to the relationship between MDC and FTC as "a can of worms", indicating a lack of faith, distrust and a perception of poor delivery by the district council. In terms of IfF's relationship to the then Conservative-run MDC, one Sustainable Frome spin-off group went so far as to describe the relationship between Frome and MDC as "enemies".

MDC governance was also felt to be highly guarded and restrictive. One respondent cited an example of when some MDC officials attended a meeting at the Cheese and Grain and only pre-agreed questions could be asked; apparently, they also bought a security guard with them to the event. Some respondents described MDC governance entities or role holders as "incompetent", "dysfunctional", "incredibly slow", "of no use whatsoever", "shady" and "not transparent", reflecting the distrust and cultural differences between these groups (such perceived identity disparities expressed in interviews with various actors are characterised in Table 6.1, below).[6]

There were clear boundaries on the salience of public roles for Frome-based actors, particularly in planning or the district council more

[6] For a summary of which types of actors I interviewed that informed perceptions outlined in Table 6.1, please refer to Table 5.3.

Table 6.1 Descriptors of the niche and the regime across entities in Frome's NDP and placemaking networks

	'Regime'				*Niche/regime interface*	*Niche*		*Place*
Dimension of power	*MDC (general/ elected councillors)*	*MDC (LPA)*	*Developers*	*NDP process*	*FTC*	*IfF*	*Sustainable Frome*	*Frome*
Governance	Turgid, reactive, slow, changing the rules, hierarchical. Stuck within the "machinations of local government".	Insufficient resources, uses 'tick box' approach to planning.	"Sewn up" resource allocation.	Centralised, slow/ protracted, bureaucratic, ordinary. Authority channelled to the Secretary of State and the Examiner, reliant on volunteers, encourages 'piggy in the middle'.	Orchestration of transition governance, standing orders, external and internal committees (tied to regulatory conditions); hierarchical, operates within three-tier administrative political system.	Enrol expertise for relocalisation, diversity, brotherhood/ clan, horizontal structures.	Enables spin-off groups. "Talking shop", Horizontal structures.	Informal. multiple membership.
Culture	Don't drive things, aggressive, inflexible, risk-averse, Conservative.	Firefighting, dealing with "hassles", time constraints, unhelpful, not proactive, not progressive, encourages 'competitive' development (national policy).	Aggressive, encourages "battles", opportunistic.	"Council state", depressing, unable to maintain momentum, leads to disappointment/ depression.	Encouraging pro-active behaviour, strategic, enabler, radical, seeking self-sufficiency/ to be relocalised. Organisational values seep into role holders (varying).	Progressive, persistent, experimental, diverse, respectful, radical.	Purposeful, alternative spiritual practices, experimental, cultivates confidence, counter-cultural, radical.	Small and Medium-Sized Eterprises (SMEs) and creative industries core to local economy, Green 'DNA', attracts 'nice people', historic core, historic town centre and market town.

(*continued*)

Table 6.1 (continued)

	'Regime'				*Niche/regime interface*	*Niche*		*Place*
Skills and interests, e.g., human capital	Incompetent, only have regime insights—not innovative or able to think creatively.	Not consider aesthetics of development, lack of experience outside planning.	Legal, planning, profit-driven.	Technical, procedural, loss of local knowledge through role change.	Highly skilled.	Creative, counter cultural insights, regime insights. Unexciting.	Knowledgeable on climate and community-led issues.	Creative, artistic, something to be 'understood', capable people.
Networks, e.g., social capital	Disconnected from Frome, lack of ties, formal ties with local levels, positive internal ties, low connection to environmental groups.	Divorced from New Economy, progressive ideas and actors.	Undercuts community-led development.	Rupture in local-LPA ties. Lack of continuity. Children and minorities unengaged.	Locally rooted ties, ties to high levels of government/ policy spaces, building relations with "good" developers.	Strong local networks, informality, ties to Sustainable Frome and Transition, multiple membership. Narrow conduits. "Pick up the phone" relationship with relocalised network.	Friendship, multiple membership. Relative independence from the Transition Network/ wider *Transition* movement.	Tucked away, drawn to action by a love of Frome. Identity re-created by the usual suspects.

Agency	Institutionalised, “politically restricted”, lock-in and curtailed autonomy, discourages self-starters.	Directed to prevent, obstruct, pushing others back.	Reactive to policy changes, resisting and challenging NDP policies.	Re-scripted intentions, role boundaries compromised “reigned in”, “tied hands”. Narrow, restricted.	Strong agency (esp. males), regime conditioning persists, power to challenge development concepts.	Pushing the boundaries, hard to dislodge “hold their own”. Free range.	Interspersed, transcendental. Energy reduced when key players left. Co-opted.	Sorting agency of place to determine the “right kind of people” to live in Frome.

generally. For instance, "I, I don't want to work in local authority planning, 'coz I think I would probably die of boredom" (IfF councillor) and "I could never work in a local authority" (Local group representative). Thus the normative, cultural values of some Frome residents—often considering themselves as protagonists—were asynchronous with any formal 'regime' identity of the state.

Condition 4. IfF had a unique blend of Transition-influenced counter-culturalism mixed with regime insights. The unique blend between niche and regime insights allowed for a cocktail of strategy and subversive action once IfF came to power. IfF had clear links to Transition through Macfadyen as well as via some other earlier councillors who were members of Sustainable Frome (Table 5.1). The notion of parallel infrastructure, which is a clear identity marker of the *Transition* movement, has been a fundamental influence on IfF's politics. This is overtly stated in a post *Our 10 Tips for Reinvigorating Local Democracy—in one place* by Macfadyen on the Transition Network's website:

> *So what we need is Transition Towns to be running parallel systems which effectively come up and swamp the so-called democracy and make it irrelevant.* (Macfadyen, 2014a)

These *tips* highlight the strong correlation between IfF and Transition and where the fusion of key identities and practices may be dispersed between the two models of transition and local democratic practices (through Macfadyen)—i.e. the 'tribes' under study, see Table 5.2—although with a slightly more politicised edge than may typically accepted within *Transition* (see Fig. 4.1).[7] Aside from Macfadyen, another key pioneer in IfF, Mel Usher, had worked as a Chief Executive for a county council and had deep insights into local government that were used by IfF as the foundations for their active dismantling of FTC's internal rules and decision-making. Echoing the notion of prefiguration, IfF's approach was to promote "change from within", "firstly taking power, then, as quickly as possible, changing the basic systems to make them more open and engaging" (Macfadyen, 2014b, pp. 11–12). At the same time, this was achieved by strategically closing off roles to incumbents that sought to

[7] These were (1) celebrate, (2) participate, (3) engage, (4) big issues of sustainability, (5) ambition and connection, (6) food, (7) tipping points in common sense, (8) new parallel systems, (9) laughter and (10) revolution (Macfadyen, 2014a).

protect the town council from the new political upstarts when it was clear they couldn't be "won over":

> *[T]he first thing [one FTC officer] said to us was 'My job is to protect Frome from you lot'... we thought we could win [them] over ... so, let's be nice and take him to have lots of drinks. Didn't work at all, it made [them] worse... it's perfectly legal but we made [their] post redundant.* (IfF councillor)

The Rolling Stone of Political Innovation

Incomers and demographic change can lead to propensities for left leaning, 'progressive' (political) values. 'New Frome' (i.e., incomers to the town, anecdotally believed to mainly be coming from South East England and London in particular) were considered by one IfF councillor to have an underlying "sense of left, trendy, artistic, green interest". With "more and more people coming", such that this changing demographic was thought by one Sustainable Frome member to be like a "rolling stone", these incomers had an influence on the voting base for IfF. The increasing profile of this demographic was considered to have affected local electoral preferences (i.e., of greener, left-leaning persuasions) to "maybe over 50% in the town", in one respondent's view. Whether this group is more inclined to vote than those typically disillusioned with politics is another avenue for investigation, but it indicates that IfF, if this stat is to be believed, had a good chance of winning the votes of most residents (see Chap. 9 for more discussion on the implications on democracy and placemaking arising from demographic change).

An 'I'm for Frome' town-focused approach transcended ordinary voting preferences (at the town council election at least). IfF councillors I interviewed recounted how even those who were not considered 'New Frome' were prepared to separate their party-political affiliations in *local* elections because of the number of positive initiatives IfF had created in its previous administration, including the benefits of solar panel installations across the town and the income it generated through the Feed-in-Tariff. This is acutely reflected in the 2015 election campaign, which cultivated support for local organisations to build shared relocalised initiatives under the banner '*I'm for Frome*' in a campaign video IfF created recounted the impressive number of changes FTC had realised since 2011 when they gained a majority (successes sung by an one IfF councillor from the first administration, a musican Al O'Kane—again reflecting the fusion of creative and political capital).

By enacting practical projects, FTC sought to overcome party political loyalties and reframe politics into practical projects that resonated with existing political identities across the party spectrum: the economic self-sufficiency of Conservative party policy as well as the 'loony left'-leaning socialist policy related to social justice through Fair Frome (a group supported by the town council to focus on issues of affordable housing, equity and inclusion). In this way, material gains in socially inclusive and sustainable initiatives helped to circumvent political identities (The effect of the rise of 'just transitions' and well-being discourse across the political spectrums would also be a fruitful study examining the role of more 'left' discourse on wider voting habits and political identities).

Notably, IfF used its Ways of Working (WoW)—agreed norms that councillors agreed to align their decision-making—to formalise a 'Frome first' approach. However, when I attended an event with IfF councillors, there was doubt among some IfF councillors just how significant these were in day-to-day operations. One councillor suggested that they were made and then put to one side, while another suggested that the reason why they were put to one side is *because* they allowed the groundwork of effective relationships to be established at the outset, reducing conflict and unclear role boundaries. However, as we shall see, even within such a 'noble' set of values, the engagement within groups, such as IfF or Transition, and especially between actors of 'difference' is inherently political—and therefore contested (see Mouffe, 1999).

Being (politically) innovative attracts high-calibre staff. As IfF's (non-)political culture became more established, this attracted qualified individuals to work in FTC roles, some of whom had experience in more senior roles at county councils or international Non Governmental Organisations (NGOs). FTC under IfF is unusual in that it has a high number of paid staff (18 at the time of the Development of Frome's Neighbourhood Plan being adopted) and included a Regeneration Manager, a Planning Administrator and a Resilience Officer and has since expanded even more. Thus, indicating how the culture of FTC was attractive and mirrored highly skilled workers' ambitions. Indeed, it was the strength of the organisational culture of IfF and FTC—and the existing skills and expertise of staff and councillors—that encouraged senior role occupiers to later join FTC. Thus, the organisational personality fit was an important factor in enrolling new actors to work for FTC that replicated the in-group (organisational) norms and affective ties (i.e. belonging and identity) that had partly stemmed from earlier networks emanating from Sustainable Frome (and the blended knowledge of regime insights of local government at higher tiers).

Reciprocity and Informality as a Tool of Governance

Utilising informality as a tool of town council governance. One of the shared practices that transferred from Sustainable Frome to IfF's own organisational practices was the informal nature of meetings (decision-making procedures)—principally because the personal nature of some of these shared ties (as friends or acquaintances) transcended *institutional* ties (e.g., being a councillor or a particular political persuasion). This enabled transformative role practices, such as the sharing of food—"distributing home-made biscuits" (Macfadyen, 2014b, p. 75)—before meetings and the use of first names amongst IfF councillors and the citizens who attended formal town council meetings. Thus, helping to break down the image, or identity, of the role of FTC within the space of public and private *belonging*. For instance, it was no longer required for people to stand before the mayor and to have only three minutes in which to speak. IfF attempted to create a more relaxed political setting. Here, pre-existing in-group ties helped to condition the normative rules and structures of FTC under IfF, which helped to facilitate a transfer of organisational identity between Sustainable Frome and IfF; indeed, one councillor accidentally conflated the two groups on more than one occasion.

Reciprocity is the key to supportive networks. The underlying character of Frome residents and a sense of intra-group solidarity were perceived as "fertile ground" for the type of politics IfF was proposing. The vision for Frome by one active local entrepreneur as a 21st market town (a model that proposed to re-work the economic model of market towns to reflect Frome's creative identity) was shared by FTC. The council wanted to future-proof the local economy and enable creative industries to thrive in the town, drawing on its historic roots and functions to reverse patterns of out-commuting and encourage local economic multiplier effects. Planning policy was felt to have gone "completely against" this and these objectives acted as a backdrop to the brief for the consultants FTC employed to support the development of Frome's Neighbourhood Plan.

Initiatives that can be seen and experienced can help to generate momentum and community acceptance of change. The notion of *momentum* to create an appetite for change was mentioned by a number of respondents, which, if carried out on an incremental basis, was felt to help people transition from fear to acceptance, such as seeing the growth of the Frome Independent Market as a local asset and part of the town's identity as a market town. Moreover, the initiatives of the Frome Renewable Energy Coop (FRECo) to install solar panels on the roof of Frome Football Club was seen as a means to bring sustainable energy into the fold of Frome's spatial setting. By proactively using government Feed-in-Tariff (FiT) incentives as a cross-integrative approach to transition, this helped to reach beyond those inclined to join green social movements into the everyday identities of the town (such as local football club fans)—or a group that might be characterised locally as "Old Frome".

Policy Alignment Between Local Government and Its Influence on Innovation

Political alignment to be innovative is heavily directed and constrained by the links between levels of local government. Politics more generally was felt to be disconnected between local and other levels of government, which IfF was seeking to address through a search for a more devolved polity and resisting attempts by higher tiers of government to be "browbeat" by them.

> *We realised that despite all the rhetoric about [the] Big Society*[8] *and empowering neighbourhoods, about them controlling their own development areas… We were hugely constrained by other tiers of authority… So actually, what we can do is quite limited by that. And the politics between Frome and the District, who are the plan[ning]—housing authority—are very different.* (Consultant)

[8] The Big Society was a government discursive framing of the need to empower communities and civil society to be more active in shaping where they live and it was a precursor to the Localism Act 2011. See 'Communities are ready for the Big Society' https://www.gov.uk/government/news/communities-are-ready-for-the-big-society.

> *[L]ocal democracy doesn't work, across the country, the layers above can't work because there's no way that they can—where do they get their information? Because if you've lost touch with people's views and needs and all the rest [of it]—then as you work your way up, how can they possibly function? Because they've got no way of knowing what's happening, even between district and town—I don't think. Unless you've got a vibrant, thriving engagement at a local level.* (IfF councillor)

The mantra of "still only a town council" was repeated several times by respondents, indicating FTC's lack of formal powers, particularly around planning policy—which massively structures placeshaping outcomes—despite its best efforts to act autonomously:

> *[W]e're not the planning authority [so] there is still that [pause] problem. But I think...what we've got to continue to do is kind of push the boundaries of what the town council does. And push Mendip... I do think we've enlivened [elongates word] local politics... I think to make it interesting, you've gotta keep pushing. Because [sighs, laughs] sounds really big-headed, but we have much better councillors than they do at the district level. Much more creative, I mean partly because it's Frome-based, erm, better informed, better, er, more imaginative about—and not just going through the motions of the status quo, which is so turgid and erm, kind of reactionary in every sense. In that, well, reactive you know, it doesn't [intake of breath]—they don't really drive things I feel, at, at Mendip District Council.* (IfF councillor)

Moreover, LPAs were considered to use a "tick box" approach to engagement with lower tiers of government, which one IfF councillor described as leading to a sense of "exhaustion...because of being pushed back so many times... We may be radical, but actually we ought to be normal". Only a sea change in government—"a radical shake-up in principle authorities"—was perceived to facilitate an improvement in FTC–MDC relations by two Frome-based respondents (an administrative rupture that did indeed happen when MDC was disbanded during the creation of Somerset Council in April 2023, which we will explore in Chap. 11.

Ironically, during my interviews in 2015, the potential for more disruptive political change through another referendum was not anticipated; the loss of the Scottish referendum was considered to mean the boat had been missed for wider political and constitutional reform on how the UK's administrative functions operate and relate to one another. Little did they know the Brexit referendum was just around the corner in June 2016.

Local innovation as a stimulant of a social movement action. The experimental political approach of IfF and their success in winning all 17 seats was also encouraging organisations and networks to emerge both in Frome and beyond—which we will discuss more in Chap. 11—Government and Politics—Changes in Representation in Political Transition Arenas. The artefacts of Peter Macfadyen's two books Flatpack Democracy: a DIY guide to creating independent politics (Macfadyen, 2014b) and Flatpack Democracy 2.0: Power Tools for Reclaiming Local Politics (Macfadyen, 2019) also did much to structure people's excitement over what might happen if they overhauled their own local democratic systems in their town and parish councils, and beyond. Macfadyen's ties to counter-cultural challenge also helped to build a movement through which to raise the bar of the politically possible at a local level, the potential of which we turn to in the later chapters of this book (notably Chap. 11).

Conclusion

Here we have discussed the role of more informal, personal ties on the politics of transition, cultivated through Frome's vehemently strong relocalised placemaking networks that permeated the town. We have seen how these ties now entered into the fabric of local government through former Sustainable Frome members becoming councillors or FTC officers. This cross-over of roles between environmental organisations and local councils helped to give rise to emergent conditions that allowed IfF and to consolidate power in its first administration and subsequent local elections. The perceived cultural differences between Frome-based and MDC-based roles indicates the high degree of othering and difference that characterised the setting of a tussle between 'niche' protagonists in Frome and the 'regime', mainstream district council which set the tone for role conflict by protagonists in the newly IfF-controlled town council and its interlocking, relocalised placemaking networks. Next, we assess how these protagonists used the planning process as a tool to advance and institutionalise their goals sustainable goals even further and the challenges associated with doing so.

References

Macfadyen, P. (2014a). Our 10 tips for reinvigorating local democracy—In one place. https://transitionnetwork.org/news-and-blog/our-10-tips-for-reinvigorating-local-democracy-in-one-place/

Macfadyen, P. (2014b). *Flatpack Democracy: A DIY guide to creating independent politics*. Eco-Logic Books.

Macfadyen, P. (2019). *Flatpack Democracy 2.0: Power Tools for Reclaiming Local Politics*. Bath: Eco-logic books.

MDC. (2005). Planning brief: Saxonvale. Retrieved from Shepton Mallet.

Mouffe, C. (1999). Deliberative democracy or agonistic pluralism? *Prospects for Democracy, 66*(3), 745–758.

ONS. (2011). *Frome census data from HMG*. Office of National Statistics.

CHAPTER 7

Using Planning as a Tool for Subversive Localism: Ambition Versus Rules in Localism Transition Arenas

Conflict Between Relocalised Approaches and the Planning Regime

Relocalised visions conflict with sterile notions of "build and bugger off" mainstream development approaches. The critical role of planning and the acquisition of a site to support development alternatives was suggested by several New Economy actors. As a member of one such organisation suggested, innovative development and design models have the potential, "to … bring together people across the political spectrum to reimagine their local economies and use those tools to do something really astonishing." Yet, there were clear tensions between a relocalised vision for development and the "sterile" structure of the mainstream development industry (which delivers most housing in the UK)—Archer and Cole (2016) estimated that by 2015, the top ten housebuilders were accountable for 47% of completed residential development, compared to only 8–9% in the 1960s (Wellings, 2006). This degree of homogenisation and lack of a locally-led development model was seen to have no long-term investment in a place, which, as one consultant I interviewed suggested, "alienates people" and went against Frome's historic character.

As a result of planning and the spatial configuration of housing estates in the north-east of the town, designed to support commuters' housing needs for those travelling to neighbouring Bath and Bristol in the 1970s, Frome

A. Burnett, *The Politics of Transition*, Palgrave Studies in Sub-National Governance,
https://doi.org/10.1007/978-3-031-40769-7_7

was broadly referred to by many as a "dormitory town"; Bath, Bristol and neighbouring Warminster were felt to have a "pull" on local employment patterns, contrasting with more relocalised visions of development.

There was a recurrent feeling by relocalised actors to create more "positive development" that demonstrated new development models that mirrored what more than one respondent referred to as Frome's green "DNA" and a perceived broad support and interest in developing alternatives, such as cohousing and self-build.[1] Saxonvale, the mainly derelict site in the middle of Frome (4.76 ha)[2] has become something of a battleground over what constitutes the "right" kind of development in Frome. For many respondents, Saxonvale represented an opportunity to create a positive relocalised vision for the town and to question the assumption that a supermarket, or similar 'mainstream' development response, is a default preferred regeneration option—with some suggesting the supermarket model was "finished" (though the rise of the more budget supermarkets, such as Lidl and Aldi make this claim uncertain—particularly in the context of the current cost-of-living crisis). It also facilitated a new approach to community-led development to redistribute local social and economic value. Saxonvale was widely perceived in the town as a beacon of hope for relocalised development outcomes. As one consultant suggested:

> *[W]hat we need to do in Frome is approach it in a different way... I'd love to see this development funded by local people [laughs]... And [they] could then share in the profits of that development, all the income that derives from that development over time. Why do we need financers and developers to do that sort of thing?" ...by excluding the key landownerships of the big owners, we start to change the economics of the site.* (Consultant)

> *[I]f we can't save Saxonvale then I fear for us, because that is such a wonderful opportunity to do something very different.* (Consultant)

Yet multiple landownerships were considered to impede changes to existing development practice whereby planning applications were not

[1] Though the term 'DNA' is in itself a contested term since it implies something quite static and structured (with certain latent states activated by environmental conditions). Importantly, and while there is an overtly green and alternative identity to Frome there are many in the town who do not necessarily ascribe to these specific identities, such as the blanket categorisation of "Old Frome", perceived as typically working-class residents.

[2] Local photographer Tim Gander has captured the changing face of Saxonvale of the online portfolio entitled "What Happened Here", available at https://www.takeagander.co.uk/-/galleries/what-happened-here (last accessed 11.6.23).

considered holistically due to differences in landowner interests (profit-driven vs. Frome-focused) that blocked mutual agreement, rather than any concrete issues with the site itself. One Frome-based group member felt that one particular landowner was "holding the town to ransom" with its "unrealistic" and "obscene" expectations for how much profit they could make, irrespective of what type of development should ensue to benefit the town more generally (i.e., well-being, design, low-carbon materials and energy, etc.).

Planning is a battlefield between different metrics to evaluate the function of place. One local authority representative suggested that planning was deemed a subjective (transition) arena which encouraged people to sit on different sides of the fence (e.g., Frome for All vs. Keep Frome Local) creating both "winners and losers". There was a sense that engagement with planning policy was a "battle" and something to be "fought", which caused one respondent to almost give up on their relocalised objectives for the town due to disaffection with the planning system. Some developers with a more aggressive approach built into their business model had a reputation for "pushing the boundaries" (interesting, a phrase also applied to the Independents for Frome (IfF) by those I interviewed). However, more heavy-handed tactics to railroad plans through the planning process were considered by one local authority representative as being "counterproductive" in the long run; this leads to a sense of alienation of communities caused by a constant challenge and acute adversarialism within the planning process which is felt in people's everyday experience of living in a particular place.

Local Planning Authorities (LPAs) are perceived as cautious, slow-moving or "dinosaurs" in their approach to placemaking. There was a sense of the LPA being "old fashioned" in their approach to employment and not being able to understand more innovative approaches to placemaking or for planners to be more strategic in their determination of applications (see Table 6.1). Despite Forward Space (Frome's first co-working space) proving a success as a space to corral creative enterprises that "came out of the woodwork" when it was set up, town-centre based economies were compromised by the planning process, unable to grow to meet demand which as one local group representative suggested, "bleeds out" to nearby towns and cities. The threat of legal challenge by developers against Local Plans in an age of austerity and a scarcity of resources was considered by a local authority representative to favour "better resourced" developers who can afford to use this mechanism to influence planning in their own interests, which influenced a more cautious approach to determining applications.

The LPA at the time, Mendip District Council (MDC)—which was superseded by the new unitary authority Somerset Council in 2023—was perceived to have powers they were not invoking that could prevent more 'mainstream' planning applications from being granted, for example, a supermarket on Saxonvale (see Chap. 8 for more discussion on Saxonvale). For instance, those I interviewed felt MDC could potentially have compulsory purchased other parcels of land on the site (whose ownership was a complex mix of MDC, a pension fund, private investors and eventually Frome Town Council, FTC). This would have enacted the power a local authority has to purchase land without the consent of the owner, with a view to bringing forward a particular type of development, such as regeneration schemes. Other powers available to MDC as the local authority included the ability to issue local development orders (which offer more flexible conditions for development on specific sites, such as removing the need for planning permission and allowing a scheme to be built out as permitted development, under certain conditions) or the local authority could have refused to sell their land to any application that involved building a supermarket. The fact that none of these options were used led to a feeling of inaction, or going along with traditional forms of development, and one local business owner suggested that the 'Saxonvale question' had become like "Groundhog Day", with nothing ever changing to unlock alternative development models. Meanwhile, the protracted nature of the planning process—and the fact these additional powers were not being used to speed the process up on Saxonvale—was felt to have frustrated and hampered Frome's ability to meet the demand for relocalised businesses in the town.

Planning is (generally) not sufficiently agile enough to capture and store the creative and dynamic energies of communities. One respondent indicated their experiences of planning in other local authorities hampered the potential to "capture and store" creative (transformative) energies from purposeful events. For example, festivals—where an "incredible dynamic, the sense of possibility and co-creativity" that was felt to be transformative for forging new bonds and a sense of purpose was often temporarily created. Planning was felt to hamper both the demonstration and the realisation of sustainability principles in practice, for example, using sustainable materials in an experimental capacity. Though, these experiences by relocalised actors contrasted with the view that the point of planning was to "get in the way" by a national-level respondent (because of its role in protecting local assets such as heritage, the natural environment, employment sites or community facilities).

Top-down housing targets conflict with a perceived capacity of place to absorb sustainable population 'growth'. An IfF councillor I interviewed suggested that Frome was considered as a "growth town" by MDC in the Local Plan and there was significant contention around the level of development the town was expected to absorb. One IfF councillor jokingly recounted that: "we've got more development in Frome than we can keep an eye on. I mean, we're full really (laughs)". For instance, Frome was expected to absorb 25% of the former Mendip district's housing requirement (out of a total of 9635 new homes between 2006 and 2029, MDC, 2014, p. 40). Since this original interview, the population of Frome has grown by over 2,000 residents over the inter-census period 2011 and 2021—indicating both the popularity of the town but also the way in which planning impacts on the contours of place by changing the built environment to absorb population growth.

Perverse Metrics and the Assessment of the Value of Development

Financial incentives to encourage innovative development and place-making are often perverse. In the words of a local group representative, the development industry was seen as based on assumptions that were "too formulaic", ignoring nuanced well-being metrics or not considering broader ownership and occupation issues or based on perverse "hope value" of future gains of landbanking. Additionally, the Local Plan was felt to be "incoherent" and contradictory regarding simultaneous employment and housing provision. Decisions on design, type (i.e., use class) and the siting of employment spaces by the LPA were felt by many Frome-based actors I interviewed to contradict relocalised economies. The latter became harder to achieve as more employment units were pushed out of the centre of the town, which many respondents felt invariably overlook the needs of the artisan economy so central to Frome's historic and current identity as a vibrant market town. One IfF councillor suggested that there was a sense that the former MDC was paying "lip service" to low-carbon development and seeking to maximise the level of development, to "prop up" their budgets from gain arising from the New Homes Bonus (a grant paid by central government to incentivise local authorities to stimulate the provision of housing).

A policy mechanism as a tool to restrict human desire for growth was felt necessary within a context of promoting reciprocity and collectivism to address behavioural defects within planning's socio-technical system.

Frome's Neighbourhood Plan—A Vehicle for Community-Led Transformation?

One way in which local actors sought to influence local development was through Frome's Neighbourhood Development Plan (NDP), which sought to respond to various development challenges in the town to advance relocalised development alternatives, encouraged by the promise (tools) of localism. Frome's NDP became a central part of how, under IfF, the ideas of relocalisation and strong narratives of independence circulated within and between social movements, such as Sustainable Frome, Frome-based residents and businesses, as well as how the plan affected relationships with formal institutions (i.e., MDC as the LPA). However, previous engagement with planning actors and processes within the planning regime illustrate the challenges of influencing its rules of engagement. and the challenges and opportunities in doing so. Looking critically at Frome's Neighbourhood Planning process reveals other important insights or conditions into the context of community-led planning in action under localism.

Intentions of Frome's Neighbourhood Plan

Frome's Neighbourhood Plan (Fig. 7.1) was 'made' (adopted) in October 2016 after a local referendum where the plan was endorsed with 3112 votes in favour of the plan and 509 against, with a turnout of 17.9% of the vote.[3] The plan is rooted in the principles of relocalisation which has the overall aim:

> *To build a community that is resilient in its capacity to support the needs of residents in the face of global shocks such as economic downturns, rising energy prices and climate change.* (Made Frome NDP—FTC, 2016, p. 3)

[3] MDC Legal Services stated that it was not possible for Frome's NDP to be held at the same time as the elections as it was already a triple election (local, district and national) and the NDP vote might get "lost" in wider political issues (according to a local authority representative). Indeed, the adverse effect on voting preferences for independent candidates when there are triple elections on local issues was corroborated by another national-level respondent who suggested that independent local candidates lost seats overall for the first time in five years in the 2015 elections because people tend to prefer to align their party-political vote when faced with a voting choice across several tiers of government (national independent political actor).

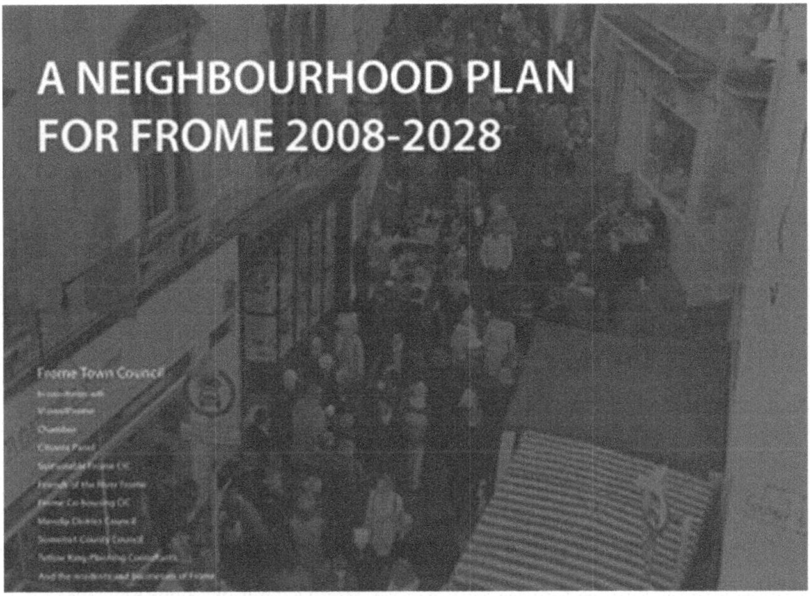

Fig. 7.1 Frome's Neighbourhood Plan front cover

The plan had several objectives to promote (relocalised) sustainable development, which are summarised in Table 7.1.[4]

The NDP includes strongly relocalised ideas and practices on community housing (community-led development), self-build (building one's own house, not developer-driven) and shared space schemes (re-prioritising pedestrian access over cars). The plan also sought to include a general policy to apply One Planet Living principles (OPL, i.e., living within the resources of one planet, Fig. 7.2) to local planning decisions, which subsequently became an aspiration—or non-plan-making approaches (CSE,

[4] Note the discursive reframing of 'golden threads' to be orientated towards sustainability, community and participation compared to the original NPPF (DCLG, 2012) where a 'golden thread' is construed in the context of a presumption in favour of sustainable development (para, 14).

Table 7.1 Frome's NDP objectives and related policies

Objective	*Related policies and/or sustainability objective*
Encouraging small businesses to start and grow.	BE1—Protection of Employment Land
Enabling people to build their own homes, live more affordably in a more energy efficient way and play a more active part in the community.	H1—Building a Balanced Community H3—Self Build and Community Housing Business and Employment Sustainable Objective 5 Business and Employment 'Achieved by' Community Energy
Making it easier for people to get around the town on foot and by bicycle.	T1—Integrated Transport Strategy
Regenerating the town centre so that it functions better as a place to live, work, meet, shop and visit.	TC1—Town Centre Remodelling TC2—Westway Centre TC3—The Cattle Market Car Park
Making the most of the River Frome corridor as an environmental and recreational asset.	POS1—The River Corridor
Ensuring future development does not damage Frome's unique character.	H2—Building by Design D1—Design in Urban Landscapes D2—Gateway Site Improvements D3—Visual Impact on Skyline

2015).[5] As we will see, these OPL principles were not permitted to be a policy per se in the final version of Frome's Neighbourhood Plan as they did not accord to the need for planning policies to be clearly able to influence the decision-making of planners to help them decide the conditions of specific planning applications). The plan also sought to protect existing employment sites, central as they are to the relocalised agenda. (Though, it is noted that a policy of substituting employment sites lost to residential development was not included in the final version due to advice from the LPA). For a summary of policies in the final plan (comparing original intentions), see Appendix D—Made policies in Frome's Neighbourhood Plan (2016) and Appendix E—Deleted policies in Frome's Neighbourhood Plan (2016).

[5] For other cities formally adopting the OPL framework see https://www.bioregional.com/projects-and-services/influencing-wider-change/one-planet-cities (last accessed 24.4.24). These include Bristol, Brighton, Peterborough and the London borough of Sutton.

Health and happiness	Encouraging active, social, meaningful lives to promote good health and wellbeing
Equity and local economy	Creating safe, equitable places to live and work which support local prosperity and international fair trade
Culture and community	Nurturing local identity and heritage, empowering communities and promoting a culture of sustainable living
Land and nature	Protecting and restoring land for the benefit of people and wildlife
Sustainable water	Using water efficiently, protecting local water resources and reducing flooding and drought
Local and sustainable food	Promoting sustainable humane farming and healthy diets high in local, seasonal organic food and vegetable protein
Travel and transport	Reducing the need to travel, encouraging walking, cycling and low carbon transport
Materials and products	Using materials from sustainable sources and promoting products which help people reduce consumption.
Zero waste	Reducing consumption, re-using and recycling to achieve zero waste and zero pollution
Zero carbon energy	Making buildings and manufacturing energy efficient and supplying all energy with renewables

Fig. 7.2 One Planet Living Principles (Bioregional, n.d.)

Above all, NDPs must demonstrate general conformity with the LPA's Local Plan and extant national legislation (Box 3.1). While the local authority respondents also indicated they sympathised or supported the "intentions" of the Neighbourhood Plan and spoke very highly of Frome in many regards, the experience of respondents involved in the plan was that the more "radical" ideas (including the OPL principles) had been removed. Indeed, there are now plans to update Frome's NDP—almost ten years from when it was made—as some town councillors have lamented that is no specific section on sustainability in the original plan and that—for reasons we will explore below—this needs to be fixed in the next iteration of the plan, particularly so that it can adequately respond to the challenge of climate and ecological emergencies (Clarke, 2023). As such, the plan became—and continues to be—a key (transition) arena of how, under IfF, the ideas of relocalisation and strong narratives of independence circulate within and between multiple entities (Fig. 7.3).

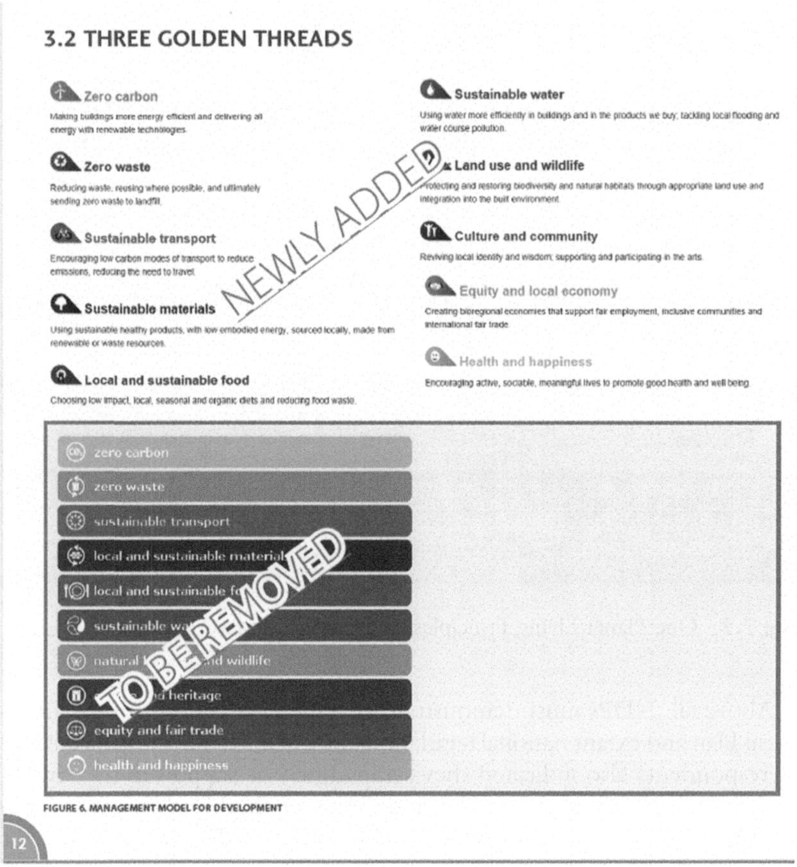

Fig. 7.3 OPL changes in the NDP v2 (FTC, 2014, p. 12). More detail was added to expand the OPL principles (NB: the older figure has wording slightly different to the Bioregional definitions in Fig. 7.2)

Governance of Frome's Neighbourhood Plan

Frome's NDP itself had several components (see Fig. 7.4), including:

- Accessing funding from government grants and FTC's own internal funds from the local precept (the tax that town and parish councils set to raise money for their activities).

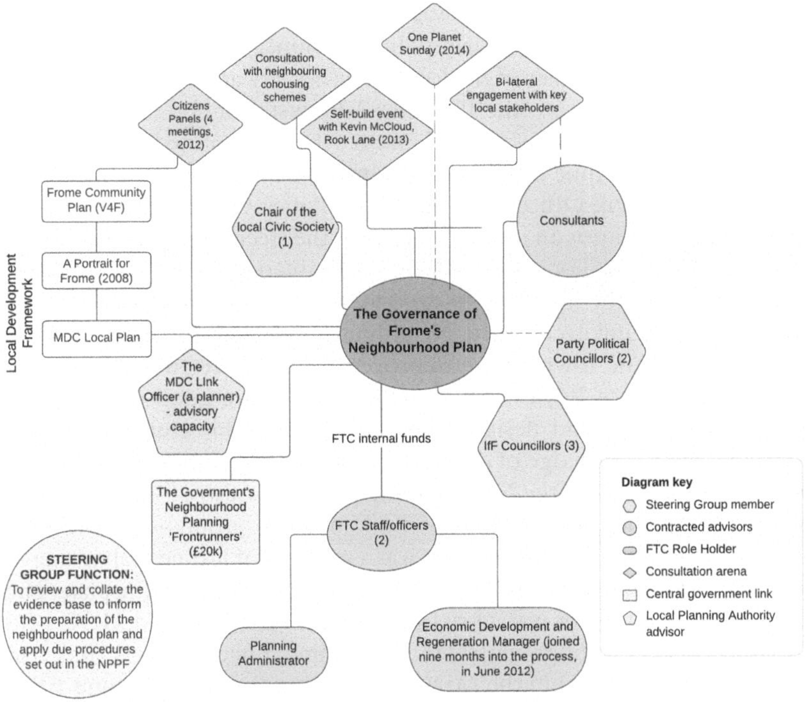

Fig. 7.4 Governance of Frome's NDP

- A steering group (to manage the process, reporting to FTC).
- Citizens Panels (to invite broader community engagement in the process, for instance to discuss whether the Community Plan—Vision for Frome (V4F)—objectives and initiatives were still relevant).
- Collation of an evidence base.[6]
- Enrolment of experts (consultants) to advise on the policy parameters of the plan, to support plan-writing, evidence base collation and community engagement.

[6]Frome's NDP and its evidence base can be found at https://www.frometowncouncil.gov.uk/council/frome-town-council/planning-for-frome/fromes-neighbourhood-plan/neighbourhood-plan-process-documents/

Frome's NDP steering group was comprised of six members, including three IfF councillors, two party-political councillors (under the 2011 administration), the Chair of the local Civic Society and was supported by two FTC staff members—the Planning Administrator and the Economic Development and Regeneration Manager (the latter joined nine months into the process in June 2012 (and who previously performed senior roles in LPAs). The Link Officer was the MDC Policy Officer who played an advisory role, acting as an interface between the steering group and MDC.

As one IfF councillor suggested, the NDP steering group was "smallish by design", with the intention to expedite the process; many of the steering group were involved in the Neighbourhood Plan, the Town Design Statement and the town centre improvement initiative (working on the pedestrianisation of Boyles Cross). The group had a good mix of planning, development and design skills—including architecture and building trades, and knowledge of different phases of the planning process. It is noted that the Chair of the NDP steering group was also a former member of the Sustainable Frome Greener Buildings sub-group—and the later chair of the Frome Civic Society also was a member of Frome Cohousing—so there were evident ties to *Transition* in the governance of the plan. The Planning Administrator was a key mediator between FTC and MDC who attended some meetings with MDC planning policy officers and had themselves worked within the MDC planning team prior to joining FTC.

NDP groups can access grant funding and technical support packages to support their plans from the government-supported Locality programme. As an NDP Frontrunner, FTC was awarded £20k to support the costs of the NDP, although the actual costs were over £60k, supplemented by FTC's own resources. Notably, these costs were used to enlist the technical and professional assistance of several consultants in the NDP process (who had a mix of formal planning and development procedure and New Economy insights). This was deemed to help advance their objectives to unlock complex development issues and represent the steering group and community's views in policy terms.

Frome's NDP sought to utilise existing consultation in the community, in the words of a Sustainable Frome member, to "short circuit" the consultation process of the Neighbourhood Plan, including the Community Plan (Vision4Frome, V4F) and other MDC-led consultation exercises. Notably, the Community Plan was believed by one local group representative to be a "solid foundation" upon which to base Frome's NDP. V4F was seen as a legitimate entry point to community engagement since it had adopted a broad and inclusive "open-ended" approach, which "intended to encapsulate as many of the community—including Old Frome—and New Frome and any other Frome wished to say. And wished to import". As we saw in Chap. 6, MDC had also carried out extensive local consultation to prepare documents that informed the preparation of the Local Development Framework, including a Portrait of Frome[7] (MDC, 2008). In the preparation of Frome's Community Plan there were synchronised and shared financial and governance arrangements between MDC and the Community Plan to support a co-dependence to feed into both the Local Plan and the Community Plan.

In the NDP consultation, several individuals in Frome's placemaking networks were consulted either through participation in the Citizens Panels or by direct engagement with FTC on core issues, for instance, employment (see FTC, 2014). The Citizens Panels were held between May and July 2012, with a total of four meetings. The panels were widely advertised throughout the town via printed and social media and the FTC website. The steering group also used random postcodes in the town in targeted priority areas to seek representative engagement of different people in the town. Attendees were also incentivised to attend the meetings with a £10 reimbursement for their time and/or expenses, though not everyone used this incentive. Despite the targeted and extensive engagement approach to bring attention to the consultation process participation ranged from 12–30 in each citizen panels (see Bourne, 2012).

The plan's steering group linked up with the well-known local TV architect, Kevin McCloud, to garner support for self-build at an event at Rook Lane Chapel in September 2013 (see Fig. 7.4).[8] The steering group also visited the nearby ecotown of Bicester and the Stroud Cohousing scheme as exemplary communities to stimulate knowledge transfer on development alternatives.

Although not explicitly linked to the NDP, FTC organised a One Planet Sunday in November 2014 to sensitise people about the concept of OPL, in part as a strategy to help garner support for the idea and to introduce an alternative framing of living sustainably.

[7] This document was part of joint 'Time to Plan' consultation and Community Plan campaigns supported by MDC and the Mendip Strategic Partnership during March–May 2008. For a summary of the documents included in Frome's Neighbourhood Plan evidence base see: https://www.frometowncouncil.gov.uk/council/frome-town-council/planning-for-frome/fromes-neighbourhood-plan/neighbourhood-plan-process-documents/

[8] http://frome.fm/2013/10/kevin-mccloud-talks-to-fromefm/

Timeline Highlights of Frome's NDP

- May and July 2012: Citizens Panels.
- 28th October to 31st December 2013: Pre-submission draft consultation (Reg 14) (which received 50 comments[9]).
- September 2014: Revised Submission draft received by MDC.
- 10th November to 23rd December 2014: Reg 16 consultation on the Submission version of the Neighbourhood Plan.
- February 2016: Examination of Frome's Neighbourhood Plan.
- October 2016: local referendum with 17.9% turnout, supported by a majority 85.7% of the vote.
- October 2016: the plan was 'made' (i.e., adopted).

It is noted that some contentious issues of the plan had not yet been reconciled between FTC and MDC at the point of formal submission of the plan, which led to an (unusual) request by MDC for the plan to be revised further before Reg 16 could take place. (The Examiner of Frome's NDP acknowledged this "modified" plan but only reviewed the Submission version for their report (Frome's NDP Examination Report—Cheesley, 2016)).

Contention and Agitation in Advancing Innovation in Locally Led Sustainability Policies

The type of policies and objectives underlying Frome's NDP were seen to be pioneering and according to one local authority representative I interviewed nothing "remotely" similar emerging from other MDC-supported neighbourhood plans. Frome's plan introduced some ideas that were not previously felt at the time to be "on the radar" of standard planning approaches. For instance, cohousing was put forward as a mechanism to deliver a more communitarian and purposeful liveability through shared living spaces that cultivated a sense of community and reduced environmental impacts which can help to approach design issues "in a very different way", according to a Sustainable Frome spin off group member. (Indeed, government policy on self-build and community housing at the

[9] See the Neighbourhood Plan's Consultation Statement (FTC, 2015).

time was becoming more favourable towards such schemes[10]). However, the idea to use exception sites—sites for affordable housing that would otherwise not normally be granted planning permission, such as those on greenfield sites outside of a permitted settlement boundary—to acquire a site for cohousing (FTC, 2014) was dropped from the plan, for reasons we will discuss below.[11] Frome-based respondents also lamented that the plan was not strong enough in challenging MDC on its stance on low-carbon standards, and that sustainable design was not pushed enough.

Key Criticisms of Frome's Neighbourhood Plan's Ability to 'Push the Boundaries' of Placemaking

The principal tension between MDC and FTC in the NDP process was in the contestation between having strong relocalised objectives that mirrored both the "DNA of Frome" and FTC's own placemaking objectives set out in its Strategy for Success (2011–2014) and the need for policies to inform planning decisions by providing clear guidance to developers. Importantly, as one local authority representative suggested, the "realities" of development control were not perceived to be aligned to the "we are Frome" mentality (itself a form of autonomous place identities) and FTC's broader agenda for wider "cuddly" ("genuine") placemaking endeavours, such encouraging OPL principles. The key criticisms of Frome's Neighbourhood Plan were varied, namely:

- Despite Saxonvale being a key sticking point for the town, Frome's Neighbourhood Plan was deemed to lack any mechanism to deliver development on the site.
- The plan did not allocate any sites for housing but instead echoed Local Plan policy and didn't go beyond this with any specific details.

[10] Though self-build and community housing were not mentioned in the 2012 NPFF, these development alternatives were in part included in the government's 2011 Housing Strategy Planning Practice Guidance which stated: "*the Government is keen to support and encourage individuals and communities who want to build their own homes, and is taking proactive steps to stimulate the growth of the self-build market*" (cited in Cheesley, 2016, para 77). Several measures were introduced including a self and custom build register and latterly the Self-Build and Custom Housebuilding Act 2015. For more information see: https://nacsba.org.uk/

[11] NDP Version 2, September 2014—Policy H6 (FTC, 2014).

However, one respondent suggested that the steering group were advised not to include sites because the Local Plan Part 2 (Allocations) was being developed at the time Frome's NDP was being agreed.[12]

- The plan's contribution to enhance Frome's open spaces was also deemed lacking.
- An idea for an eco-town early on in the plan was considered a way to cultivate more radical development approaches, including how communities could be designed with "new rules" built into the liveability of a more intentional community and replicate best practices from other countries; this idea was considered by one IfF councillor to have gone "out of the window" and disappeared into the ether (in part due to advice from the contracted planning consultants that the idea "wasn't feasible").
- FTC's intent to pursue an employment land substitution policy (i.e., if employment land were lost to residential development it should be substituted elsewhere in the town so overall there would be no net loss in employment land) was not considered able to be applied by planners in practice.
- Notably, MDC had concerns about the plan not meeting the 'basic conditions' (i.e., that the plan compiled with existing local and national policy).
- MDC suggested that OPL was not substantive enough to be a standalone policy because planning officers argued these principles could not translate into development control measures (MDC, 2015), that is, processes through which planners can check if a development is being built as agreed—and it was dropped as a policy from the plan. It became a 'golden thread' of aspirational principles, rather than being able to provide any concrete policy criteria to development in the town.
- The plan was felt to lack a "story" behind how different elements of the plan related (backed by evidence), such as defining "strong com-

[12] Part II of the Local Plan (which allocated additional housing and employment sites) was only adopted in December 2021, well beyond when Frome's Neighbourhood Plan was made, meaning there was a hiatus between the generic content of Frome's Plan, the Mendip Local Plan Part I and further detail on the allocation of sites in the district. Since it was adopted, the Mendip Local Plan Part II has been subject to various legal challenge and judicial review. See https://www.somerset.gov.uk/planning-buildings-and-land/adopted-local-plans/?district=Mendip (last accessed 12.6.23).

munities" and why a provision for 5% of the self-build policy was included in the plan.

- The idea of self-build was criticised by some working on the plan as not being a feasible approach that could work in practice nor deemed as something that came out strongly within the consultation. Also, that by fusing the principles of cohousing with self-build and community land trusts on exception sites this diluted the more radical aims of cohousing; that is, if cohousing is channelled towards rural exception sites (as affordable housing), this can compromise their economic viability (and while widening the model up to others, this could change the 'type' of person who might be attracted to such schemes if affordability becomes the main driving force of such schemes).

The Limits of the Planning Process to Encompass Relocalised Innovation

Development control logics squeeze out the "starry-eyed" elements of community-led placemaking. A number of policies were altered, or removed, by MDC (2015) and the Examiner (Cheesley, 2016). These changes were perceived as necessary to remove the more "starry-eyed" and "woolly" policies (which were perceived to indicate a "naiveté" of the steering group) and to fit these aspirations into the logics of development management (to instruct them to determine planning applications). Doing so was felt to "squeeze" the more "laudable aspirations" of the steering group because "they don't stand up to scrutiny"—words used by local authority representatives.

One FTC respondent indicated these two approaches—the 'niche' and the 'regime' logics of placemaking and planning systems—were playing "a different ball game". Ultimately, the framework and principles of OPL and other relocalised objectives were not sufficient to challenge profit-driven development models; as one consultant explained, the key to neighbourhood planning is to achieve a deliverable plan—this in effect meant "we can't put handcuffs on developers". As a consequence, much of the radical intention of the plan was left wanting because the mechanics of the local planning authority, the planning system and the Neighbourhood Plan removed momentum for alternative development schemes. As one Sustainable Frome member suggested, this meant Frome's Neighbourhood Plan turned out to be perceived as a "quite bland plan, because of all the things that had to be smoothed" because of what one IfF councillor described as the "mangle of local government".

"Get out clauses" within policies were felt to enable more alternative ideas to be included in policies, such as "in exceptional circumstances" (Policy BE1—FTC, 2016). However, these were also seen to undermine radical intentions and the strength of directing alternative development practices. One consultant indicated that innovation in NDP policy areas has varied degrees of manoeuvre, but often through a "thin corridor of innovation". Or as one consultant suggested, NDPs only allow local actors to get involved in limited policy arenas, such as design and self-build.

These challenges, of having sufficient room to innovate, led one FTC representative to suggest that ideas such as cohousing were tweaking around the edges: "When you've got policies in there that allow development at the minimum building regulations to go ahead, you're not going to save the world [laughs]. It's like fiddling while Rome burns, from an environmental perspective." Another FTC representative suggested that some development alternatives can be too idealistic or prescriptive, reflecting on Frome Cohousing's criteria to find a site for its members to prefigure their preferred development model and they suggested that the group were "looking for too much. They wanted utopia and it's just not out there".

There is a disconnect between planners supporting relocalised ideas in practice and the realities of what is permitted in development terms. Despite being an active point of contention between FTC and MDC, many of the ideas the plan sought to advance were welcomed by planners. As one local authority respondent suggested, FTC's approach warranted "all power to their elbow"—an endorsement of resistance and encouraging success in their endeavours. Yet despite this, MDC was perceived to be "unhelpful" in the NDP process and was seen to be picking faults or being wilfully obstructive, over and above the recognised constraints on resources rather than being supportive; this accentuated already "antagonistic" and "difficult relationships" (see Table 6.1).

Planning can lead to highly institutionalised role conditioning and prevent 'niche' perspectives from taking root—but levels of conditioning depend on which aspect of planning you work in. Reflecting the notion of institutional 'lock-in' described by some sustainability transitions scholars (Klitkou et al., 2015), working within an institution, such as an LPA, was perceived by many as being conditioned by an institutional regime. Planners were considered to have a disconnect between the aesthetic quality of the built environment and a very "process-driven approach", which according to an IfF councillor became more entrenched as people move "higher up the food chain" in an LPA.

Some respondents had "almost exclusively" a planning background. Indeed, "conditioning" of previous roles within government or the development industry was considered hard to break down, even for actors that had themselves transitioned into New Economy arenas. (A lack of conditioning for relocalised ideas and culture by some interlocutors was perceived as a block for Sustainable Frome engagement in the NDP process—see Chap. 8). For instance, conditioned perspectives of "viability" for actors who had worked in the development industry conflicted with the views of those without planning experience who wanted to continually push the boundaries; one New Economy actor describe those with planning experience to be "entrenched" in regime thinking and unable to "cede control".

There was a sense that the longer actors stay in regimes the harder it is to break out of institutionalised thinking. Being conditioned by a regime was felt by an IfF councillor to tame people's capacity to be a "self-starter" and encourage people to follow directions. Institutional lock-in was seen to contribute to mundane processes, where—in the words of an FTC representative—MDC officers were seen as "not able to be flexible", compounded by the fact that lack of resources meant they had to firefight rather than be proactive:

> *You know, [the steering group] got sucked into a sort of council state ... And I would say that's partly [some peoples'] background. But the whole thing did become a bit ordinary. Or more ordinary than it might have done [if FTC has had more control over the process].* (IfF councillor)

Interestingly, one respondent with a varied skill set from a career in government suggested there were variations between which aspect of the planning 'regime' one works within and their degree of flexibility; for instance, a consultant suggested that development and control planning was deemed restrictive and not pro-active, compared to designing (policy) planning briefs, which can encourage a more pro-active approach.

Relocalised intentions aren't readily aligned to mainstream logics of development control. On a related point to the one above, a local authority respondent indicated that any challenge to the underlying growth-centred discourse that underpinned localism and the planning regime was not welcome because this was the "only game in town … That's the agenda we're all ultimately working to." While another local authority respondent suggested that housing delivery targets needed to increase because of a lack of supply in recent years. Meanwhile, government policy was not allowing a "free hand" for community-led development and changes resulting from the 2015 Deregulation Act (that effectively took away a neighbourhood plan's ability to set more stringent low-carbon standards) were a "wake-up call" that indicated a more restrictive and heavy-handed approach whilst favouring developers' capacity to build their way out of a recession.

Innovative ideas were felt to not easily sit within local plans. An example of this is that whilst the cohousing group inputted into the Mendip Local Plan there was a disconnect in understanding (and therefore enacting) development alternatives. Ideas like cohousing were seen "as a bit too innovative" by one Sustainable Frome spin-off group member. A common misunderstanding of what cohousing is and an association with "hippy communes in the 1960s" was deemed to prevent such schemes from taking root. Such lack of awareness of the LPA prevented conversations and connections to be instigated. Frome Cohousing didn't find the LPA's policy team approachable on such innovative development issues; although engagement with the MDC housing policy department was being explored as an alternative entry point into the district council and working with social landlords who may be able to leverage resources for cohousing schemes.

The Limits of Neighbourhood Planning Powers

Neighbourhood Plan rights don't extend to changing the rules of the planning regime's internal control mechanisms. There was a sense that the clash of culture between MDC and FTC influenced the tone and scope of the plan, in favour of MDC's more risk-averse approach (see Table 6.1)—primarily due to the fear of appeal—which ended up being reflected in the scope of the NDP. At the same time, some respondents believed that the steering group had been "weak" in standing up to the LPA. One local authority respondent indicated frustration that NDPs were being used in a way not originally intended by the government; many communities were deemed to perceive NDPs as a mechanism to "take back control" in ways that went beyond the original scope of NDPs set by the government (interestingly, a term latterly commonly used in pro-Brexit 'Leave' debates about how leaving the EU will empower the UK to govern on its own terms). There was thus confusion among NDP groups about what is/isn't a planning issue and misusing the NDP, indicating a misunderstanding about where the thresholds between local authority discretion and community autonomy lie. This mismatch between ambition and practicalities led one local authority respondent to suggest that some NDP groups may be pursuing the wrong mechanism due to a lack of information on the purpose of NDPs.[13] At the same time, one IfF councillor suggested they had been "had" by the promise of localism.

Neighbourhood Plans take a long time and affect the levels of motivation of community-based actors. Despite anticipating a relatively rapid process, building on previous community engagement, the reality was the NDP was a lengthy process and became—in the words of one local group representative—a "fairly tortuous process" that "has crawled along", or in the words of an IfF councillor, "dragged on for so long". Another local business owner suggested that it was "just far too difficult far, far, far, too difficult" and led to local actors feeling they were "going around in circles". The plan took over three years (which is relatively quick compared to some Neighbourhood Plans) and caused both a feeling of "depression" and a loss of momentum among its steering group members. The already existing voluntary role of parish and town councillors and community organisations affected levels of engagement, where some had to choose between personal financial resilience and getting the plan across the line, highlighting how the expectation for communities to deliver on localism can be at odds with protracted and complex planning processes.

However, even within FTC, there was a sense that some role occupiers were being constrained by the regime-like qualities of *IfF*, including overriding some individuals' own capacities to exercise autonomy in their roles. One consultant described how one role was particularly challenging because they had to act like "piggy in the middle" between FTC and the steering group and MDC through the NDP process. Some respondents felt some of the steering group did not sufficiently recognise peoples' skills

[13] For instance, Neighbourhood Development Orders (NDOs) and Community Right to Build Orders (CRtBOs) were perceived by a local authority representative to be more suited for larger major regeneration schemes, not rural villages and suggested that additional guidance on what consititutes the most appropriate community planning instrument for specific purposes was needed (i.e. would a parish plan—which is simpler to produce than, say a neighbourhood plan or an NDO or CRtBO—suffice to support a community's placemaking objectives).

and expertise went underutilised. There was also an underlying sense of clarifying seniority and how some more junior staff were "managed" by others, or were sometimes keen to clarify lines of authority within FTC. It was noted that during an international exchange I attended between four (yes, four) of Frome's twin towns in 2016 on locally-driven renewable energy[14] there was no mention of Frome's NDP as a mechanism to advance low-carbon development.

Conclusion

Frome-based actors sought to counter pressures on land and the built environment by formalising relocalisation through Frome's neighbourhood plan and experimentation with other rights in the Localism Act. We have reflected on the limits and opportunities of doing so, highlighting the conflicts between relocalised visions and mainstream development approaches, the issues of valuing the importance of place through community planning and their agreement to local planning authority instruments in the local planning framework. In particular, we have looked at how innovation at a local, 'niche' level is often unsynchronised with local authority ways of organising, nor is the top-down hierarchy of planning's consultation and procedural mechanisms always conducive to harnessing the temporary and creative energies of place. This causes some of the more "cuddly", aspirational and prefigurative visions of relocalisation to be squeezed out by dominant development control logics of planning policymaking; where slow-moving, procedural matters can dampen enthusiasm to use community planning for more radical visions of placemaking in action—causing, in turn, a disconnect between planners and places. Next, we turn to examine how these constraints might be challenged through more subversive strategies to influence the underlying politics of localism between tiers of government and how this affects the empowerment of local actors to solicit transformation of approaches to influence place, community and the built environment.

[14] See https://www.discoverfrome.co.uk/international-renewable-energy-conference-in-frome/ (last accessed 19.6.23).

References

Archer, T., & Cole, I. (2016). *Profits before volume? Major housebuilders and the crisis of housing supply.* https://thinkhouse.org.uk/site/assets/files/1803/profits.pdf

Bioregonal. (n.d). *One planet living principles.* Last accessed 5.12.23. https://www.bioregional.com/one-planet-living

Bourne, L. (2012). Report and feedback from the Citizen's Panels held in Frome from May – July 2012. *Action for Market Towns.* Retrieved from https://www.frometowncouncil.gov.uk/wp-content/uploads/2014/05/frome_report_from_citizens_panels_-_final_document_july_2012.pdf

Cheesley, J. (2016). *A neighbourhood plan for Frome 2008–2028: Report by independent examiner.* https://www.frometowncouncil.gov.uk/council/frome-town-council/planning-for-frome/fromes-neighbourhood-plan/neighbourhood-plan-process-documents/

Clark, L. (2023). Frome Neighbourhood Plan to get a makeover after a decade of guiding the parish's development. In *Your Area* (accessed 5/12/23) https://www.inyourarea.co.uk/news/frome-neighbourhood-plan-to-get-a-makeover-after-a-decade-of-guiding-the-parishs-development/

CSE. (2015). *A guidebook for low carbon neighbourhood planning.* Centre for Sustainable Energy. Bristol, UK. https://www.cse.org.uk/my-community/engagement-planning/neighbourhood-plans/

DCLG. (2012). *National planning policy framework.* Department for Communities and Local Government.

FTC. (2014). *A neighbourhood plan for Frome 2008–2028 (v2).* Frome Town Council, Frome. https://www.frometowncouncil.gov.uk/council/frome-town-council/planning-for-frome/fromes-neighbourhood-plan/neighbourhood-plan-process-documents/

FTC. (2015). *NDP consultation statement.* Frome Town Council. https://www.frometowncouncil.gov.uk/council/frome-town-council/planning-for-frome/fromes-neighbourhood-plan/neighbourhood-plan-process-documents/

FTC. (2016). *A neighbourhood plan for Frome 2008–2028.* Frome Town Council. https://www.frometowncouncil.gov.uk/council/frome-town-council/planning-for-frome/fromes-neighbourhood-plan/neighbourhood-plan-process-documents/

Klitkou, A., Bolwig, S., Hansen, T., & Wessberg, N. (2015). The role of lock-in mechanisms in transition processes: The case of energy for road transport. *Environmental Innovation and Societal Transitions, 16*, 22–37. https://doi.org/10.1016/j.eist.2015.07.005

MDC. (2008). *A portrait of Frome. A background paper to help support public consultation on the preparation of the Mendip sustainable community strategy and local development framework core strategy.* Mendip District Council, Shepton Mallet.

MDC. (2014). *Mendip local plan Part I.* Mendip District Council. Now available from https://www.somerset.gov.uk/planning-buildings-and-land/adopted-local-plans/?district=Mendip#Mendip%20Local%20Plan%20Parts%201%20and%202

MDC. (2015). *Planning board: Agenda Item 7—Frome Neighbourhood Plan.* Mendip District Council, Shepton Mallet.

Wellings, F. (2006). *British housebuilders: History and analysis.* Blackwell.

CHAPTER 8

Subversive Localism?

The Importance of Positive Inter-governmental Linkages Where Local Authorities Act Positively to Enable Comprehensive Development

Localism galvanised action on community-led planning and place-making. As a set of policy and planning instruments, localism has continued to entice many with ambitions to have greater agency to make their community a better place to live. With that was a desire by some communities, such as the market town of Frome, Somerset to be bolder and more radical in how local councils shape places and the role communities play as a stakeholder in doing so. Frome-based actors also sought to imbue a strongly autonomous flavour to the opportunity that localism offered—as one local authority officer put it, localism seemed to allow "having some powers that people can seize for themselves at their own discretion". One consultant I interviewed suggested that localism had "massively radical" potential that enabled "hugely under the radar" intentions that can embrace the cross-cutting issues of democracy, governance and local action; where, they added, communities can help achieve efficiency savings and "become more resilient and come up with very bespoke solutions" to local development challenges.

In particular, the general power of competence—set out in the Localism Act 2011 to give councils the power to do anything an individual can do

A. Burnett, *The Politics of Transition*, Palgrave Studies in Sub-National Governance,
https://doi.org/10.1007/978-3-031-40769-7_8

provided this doesn't contravene existing legislation—was seen as a mechanism to transform the remit of Frome's Town Council (FTC) autonomy to take a more strategic approach to town-council facilitated placemaking. This helped to cultivate a more optimistic culture within the town to "think big" in terms of the potential impact local councils can make. Having a neighbourhood plan was seen as a strategic mechanism to enable progressive development approaches, even if IfF was no longer in power. As one local authority respondent suggested, neighbourhood plans can help to deliver a nuanced perspective that a Local Plan simply cannot achieve.

However, whether localism actually shaped decision-making within local authorities to positively respond to community-led development solutions was questioned and as one actor involved in developing an alternative scheme in Frome suggested, the extent that Frome's Neighbourhood Plan played in underpinning this scheme was almost non-existent (because of the absence of some ideas it could have latched onto in policy terms, described in Chap. 7). In fact, for reasons we discuss below one development activist in the town reflected that "Did the Localism Act ever arrive in Frome in any meaningful sense? Did it seek a re-balancing away from the targets of profit-led developers? I don't feel that it did".

Using localism rights well requires a certain degree of community-based 'maturity'. Using rights underpinned by the Localism Act was seen as a "step up" for transition actors to agitate role conflict and change dynamics, and work with wider stakeholders where developers buy-in to more sustainable and radical placemaking opportunities. However, as one respondent close to the *Transition* movement pointed out, this requires community groups to get to a "certain level" of organisational maturity to be able to "take on" planning. As one consultant put it, localism has a greater chance of working and achieving more radical intent "if the governance conditions are right". This speaks to the concerns raised by Wills (2016) and Parker et al. (2017) in Chap. 3.

Not all localism powers are created equally. As we have seen in previous chapters, Saxonvale is a former industrial and now largely redundant town centre site that has become an arena of contestation about the future of development in Frome and an existential battle over the town's identity. So far there have been five conditional contracts on the site put forward by different developers which have, in the the words of one respondent "unravelled". At one point in this complex story of landownership and contestation over who and what would be developed on the site and in parallel to the NDP process, a Frome-based group (previously established to act as an intermediary to explore development opportunities on Saxonvale) sought to use a Community Right to Build Order (CRtBO), an additional community right mentioned in the 2011 Localism Act. As one consultant described, this sought to unlock the Housing and Communities Agency (HCA) commitment to fund £100,000 "to deliver an alternative approach", with a £10,000 match funding contribution from FTC. However, not only was the CRtBO process seen as a box-ticking exercise, which made it difficult to release funding, but the former Department of Communities and Local Government (DCLG) intervened to prevent the HCA releasing the funds as Saxonvale was considered too large a site for a CRtBO, though as one consultant suggested, this effectively "pulled the plug on the funding" for the alternative scheme.

The Importance of Local Capacities: 'Strategic Intermediaries' to Unlock Protracted Development Challenges

Delivering comprehensive schemes requires a clear plan and deep engagement with the investment model underpinning development. FTC's enrolment into the nascent CRtBO process led them to follow up on an idea by a member of a local group to explore the idea of an Implementation Prospectus as a roadmap for development on Saxonvale. This would involve researching the aspirations of the site's multiple landowners and stakeholders on the conditions they had for enabling development and the delivery mechanisms to achieve them. This was set in motion by FTC and the prospectus was formally approved in March 2015 by the council. A Frome-based contracted consultant had senior planning and regeneration experience working with Local Planning Authorities (LPAs); they explored a Neighbourhood Development Order (NDO)—another community right set out in the 2011 Localism Act—as a mechanism to navigate multiple landownerships and include broader FTC placemaking objectives to facilitate the "intentions" of the policies set out in Frome's Neighbourhood Development Plan (NDP). For instance, by expanding the offer of the town centre, improving public realm objectives within Frome's Conservation Area, improving signage or using frontages to encourage social activity. One respondent close to this idea also suggested to split the site and move away from comprehensive development to fracture commercial (profit-driven) vested interests and encourage alignment of local interests and traders, circumventing commercial landowners over local or Mendip District Council (MDC)-owned land, a move that helped to move on relational ties invested in the site that eventually paved the way for an alternative scheme to emerge.

The economics of investment in the site were seen as reasons that prevented the Saxonvale site from being unlocked. In particular, a consultant proposed that government loan funding could be used for decontamination, decommissioning, servicing and infrastructure costs and funds paid back when the development makes a profit, to enable development without a high expectation of immediate profits (as shareholder models often dictate). In conjunction with other mechanisms, they also suggested that a brownfield tax could be imposed on land to encourage developers to bring it forward or to take an annuity (i.e. an annual fixed rate) rather than a capital sum (one-off payment) on profits arising from development to reverse the practice of landbanking (which seeks to achieve higher future returns for shareholders). Indeed, they suggested that any profits could be ringfenced for the community, which could help to formalise relocalisation. However, the contractual obligations of a pension scheme—one of the landowners on the Saxonvale site at the time—to maximise profit made it challenging to engage them in discussions of alternative financial delivery mechanisms, thus pointing to how the underpinning investment landscape of investors in development structures the opportunities for different financing solutions to emerge. (It is also pertinent to note that as of 2023, government investment now mirrors some of the ideas raised by the consultant back in 2014. For instance, as of 2023, the government has introduced a Brownfield, Infrastructure and Land Fund to unlock development and regeneration on brownfield sites which would otherwise be unviable, though this is focused more on 'carrot' (funding) rather than a 'stick' (tax) based' incentive structure.)

Commissioning the Implementation Prospectus for Saxonvale was felt by one FTC representative to "shake things up a little bit" and "change the view" on existing development models. Historical distrust between the town council and other tiers of government—dating back to when Regional Development Agencies (RDAs) were in place—was felt by a respondent close to shaping ideas on a Frome-based vision for development of the Saxonvale site to hamper productive discussions between landowners. Enrolling the support of MDC was considered a means to help utilise statutory powers endowed to LPAs to help change the balance of power between landowners and enable partnerships between different tiers of government (i.e. district and town councils). For instance, by being able to realise powers such as Article 4 directions that remove permitted development rights under the discretion of a local authority, meaning that development would need to be approved by the planning process. Similarly, a Section 215 notice gives a local authority the ability to require a landowner to clean up their land if it falls partly or wholly on the authority's landholdings. In addition, a partnership model between tiers of government was felt to widen the scope to access funds to unlock development. England's 36 Local Economic Partnerships (LEPs) have played a role in supporting economic development but these are being phased out from 2024 and absorbed into the wider economic regeneration duties of local authorities, which may help to strengthen economic and strategic investment in planning if aligned to wider community governance.

Relocalised development alternatives need champions who can engage with higher tiers of government. Having an experienced mediator was felt to be able to unlock historical political tensions between the two levels of government and to focus on concrete actions to help "drive" FTC goals for Saxonvale with a "little bit of a push", and in partnership with number of actors, working "under the wire" to develop an alternative approach. Whilst an NDO was not pursued—mainly due to the process having a similar level of complexity to the NDP process and the associated resource constraints of doing so—conversations with MDC arising because of this work helped to progress discussions on the way forward on Saxonvale to help find a "complementary, not a competitive role". Indeed, the presence of a supportive Conservative Member of Parliament, MP (on green and Frome-based issues) had partly mediated the relationship between FTC and MDP to reshuffle the historical impasse between these two tiers of local government and to voice new policy ideas to bolster relocalised localism at the national level; which in part may be due to the gains made by the Green party—and IfF—in the 2015 election and strengthened further in the 2019 and 2022 local elections (see Table 4.2 Key events in the study of Frome's relocalised and independent placemaking).

Localism doesn't go far enough to deliver working devolution to empower local councils. There was a feeling by some interviewees in my research that English localism didn't go far enough to realign local interests and powers to effect decisions that respond to local needs, citing European devolved models as best practice. Moreover, some people I spoke to felt that the democratic administrative system at the local level was outdated or not fully functional in terms of the accrual and disbursement of incentives to encourage particular types of development. National government policy was seen to prop up district councils without sufficient financial mechanisms to distribute benefits from housebuilding to local communities such as the allocation of funds from the New Homes Bonus which "comes in [to MDC] and then it stops".[1]

The extent that localism powers affect the conditions within which state and community actors coalesce and feed into the planning 'regime' can indicate the degree of political contestation between actors in shaping place-based design principles. Next, having seen what Frome's Neighbourhood Plan sought to achieve in the previous chapter, we now turn to reflect on how it solicited a particular politics of transition as it attempted to encourage innovative placemaking and alternative development models. Notably, how differences in political approaches between governments and the constraints of other tiers of government authority circumvented the intentions of Frome's NDP and IfF more broadly in achieving purposeful, communitarian and relocalised development (through the planning process at least).

The Politics of Planning Transitions and the Limits of Localism

The newness of localism gave rise to a lack of confidence in how to proceed. The relative newness in the early days of the localism policy space (i.e. during the time of the Government's Neighbourhood Planning 'Frontrunners' scheme, to which Frome was enrolled) was seen by some I interviewed as having contributed to a sense of learning on the job for some of those involved in preparing Frome's NDP, including consultants. For instance, as one local group member put it, in the Citizens Panels (see Chap. 7) there was a lack of clear "parameters … within which they were

[1] The New Homes Bonus is a grant paid by central government to local councils to reflect and incentivise housing growth in their areas (see: https://www.gov.uk/government/publications/new-homes-bonus-provisional-allocations-2023-to-2024 last accessed 12.6.23). It is based on the amount of extra council tax revenue raised for new-build homes, conversions and long-term empty homes brought back into use. FTC were contending that MDC would receive £10m up to 2026 and that Frome should receive its share of funds arising from new development through the New Homes Bonus (demanding £2.5m for 2500 houses between 2010 and 2026, not just a £50k legacy fund allocated to Frome). A petition to this effect was signed by Frome residents and addressed to the Chair of MDC and was previously available at http://m.mendip.gov.uk/CHttpHandler.ashx?id=2880&p=0 before the district council was abolished (last accessed 16/1/19).

supposed to be participating … and therefore, it was less effective". Moreover, this policy 'newness' also affected the confidence one consultant had to direct the Neighbourhood Plan and led to a degree of contestation in the balance of power between role expectations within the governance of Frome's Plan; they suggested: "[with hindsight] I think we could have started from a different point, and done it differently".

For localism powers to go anywhere, communities need the skills and capabilities to seize them. Echoing Wills (2016) and Gallent and Robinson (2012), the capacity issues of local communities, in general, were called into question by one IfF councillor due to the people's skills and abilities to actually deliver on localism. One particular comment of theirs indicates the broader structural 'broken links' - which are particularly informative of a study on sustainability transitions (refer to Figs. 5.1–5.3)—relating to the diffusion of the potential for (progressive) localist outcomes when powers are given to "moribund councils". In this case, the IfF councillor suggested that "giving power down to some of these people is not [to] the right people who have the influence. Because it just doesn't go anywhere basically, [power] just stops"—indicating how just because powers are available it doesn't necessarily mean they get used effectively.

Localism intensified top-down policy-making directives. There were alignment challenges between the national, district and NDP groups due to government policy changes, notably on sustainability criteria within planning policies (in particular, following the 2015 Deregulation Act). The simplification of the planning system with the introduction of the Nat ional Planning Policy Framework (NPPF) in 2012 was perceived as not substantially altering the principles of the planning regime; however, the NPPF's supplementation with additional planning guidance was felt by one planning consultant to have taken the underpinning planning policy architecture "back" and added ambiguity. Planning practice more broadly was deemed to operate on a reactive, unstable basis because of 'ill-thought out' policies which often need correcting.Indeed, this is even recognised by the government itself in a 2023 Policy Paper which states that "National planning policy has been characterised by stop-start reform over several years. This has regrettably resulted in uncertainty among local authorities and across the planning sector." (See DLUHC, 2023).

One IfF councillor suggested that at the time that Frome's NDP was being prepared, contrary to expectations, decisions on planning matters were going "up the food chain" instead of down (as promised), with some processes becoming more top-down and prescriptive, such as the number of decisions taken to the Secretary of State and the increase in recruitment at the national Planning Inspectorate (which deals with appeals, i.e. challenges by developers to planning decisions made by LPAs). Despite the promise of the general power of competence, town council officers and IfF councillors lamented that the government had given powers they've since retracted because it got "cold feet". An interesting recent policy evolution of what might be considered 'Localism2.0' can be seen in the Labour Party's 'Vision for Community Power', published in collaboration with various national supporters of community-led initiatives (Kaya Comer-Schwartz et al., 2023). This includes measures to extend participation in strategic decision-making as a new 'community right', widen the engagement of communities in spending decisions that directly affect their neighbourhood, the enhancement of social value outcomes in procurement and offering favourable conditions to support community-led or cooperative enterprises and aligns the role of planning to other mechanisms to support community wealth building.

Localism as a political strategy to appease factions of political parties? An IfF councillor suggested that localism had not been delivered in such a way to genuinely enrich democracy and had failed, more broadly, to enable this because it was seeking to "*appease* two very different kinds of Tories, kind of big business versus shire Tories" - thus pointing to the factional differences within political parties and the structuring function of identities in determining the durability of value alignment and political buy-in. There was a sense by one IfF councillor that the government was using localism and planning for their *own* subversive ends to achieve housebuilding to stimulate the economy. Moreover, there was tension between the government's priority of growth and, in the words of an IfF councillor, "delivering neighbourhood aspirations", which were not perceived as compatible.

Resource constraints hamper greater engagement between planners and community groups. A more constrained experience in austere times meant that there was a discrepancy between the amount of time MDC officers would have liked to input into the process and the actual amount of time they could contribute, with one stating "We have to cut our cloth, unfortunately … I have to say that my contact with them has been relatively infrequent really", and only at key stages of the neighbourhood plan's development.

Synchronicity and Lost Opportunities for Convergence Under Localism

The policy underpinning localism is a moving feat. As mentioned above, there was a sense among different types of respondents that national government policy discourse was to be "taken with a pinch of salt", where policies were considered vague, changeable, lacking in guidance or appealing to too many interests to be effective. FTC officers, consultants and local authority respondents spoke of their mutual frustration both with the NDP process and the broader shifting government policy landscape, or the "moving ground from under your feet". Notably of low-carbon design as a policy area that one consultant described as being "superseded" by national policy through i) the 2014 removal of the Code for Sustainable Homes (CfSH) and ii) the drive towards including sustainability issues within Building Regulations (to accommodate a blanket standard for low-carbon design under the Deregulation Act 2015). As one local authority representative suggested:

> *[Government policy can] drive the coach and horse through what you're trying to do anyway ... That really pulls the rug from any parish or other neighbourhood forum that wants to introduce something which is more draconian [intake of breath]. They just simply wouldn't be entitled to do it. And the Examiner will be bound to just put a red line through it, if it ended up in the plan. Which is a bit depressing, really.*

The effect of changes in national policy was felt to put some advisory roles in an uncomfortable position, where role boundaries and ambitions were being reigned in; these actors were forced to take a more prescriptive stance towards more ambitious NDP policies due to government policy, focusing on what *can't* be done rather than what can. A situation described by one consultant as being "reined in" and "restricted" by government policy, leading to "disappointment" among communities. (One IfF respondent suggested if they had changed consultants this negative stance of playing the role "to bring us down to earth", "to keep us grounded", to tell Frome's NDP the steering group (see Fig 7.4) "what we would go through" might have been remedied, but they were too far into the process to do so. Given the policy constraints indicated here it is questionable how much more freedom another consultant team might have had to give a more positive, and practical, spin on how far the NDP could go on sustainability issues at the time).

Lack of synchronicity to existing government policy can undermine the potential for shared governance. The alignment to the Local Plan in determining degrees of alignment with NDPs was seen as important for one consultant but less important for a local authority representative—who placed the onus of responsibility for any misalignment *onto* town and parish councils, for example, in Frome's case for initiating the NDP process before MDC policy had been adopted (which included the Local Plan Part II that allocated additional sites, which in the end took until 2021 to be approved - way beyond the timeframe of the preparation of Frome's NDP in any case). This was in marked contrast to the experience of the Community Plan, which was much more synchronised due to shared governance between MDC where their Regeneration Officer was, as one respondent close to the development of Frome's Community Plan suggested, "part and parcel of the [Community Plan] Executive team"; shared consultation exercises fed both the early phases of the Local Plan (the Local Development Framework) and the Community Plan. MDC also contributed to £25k in additional to other grant funds (bringing the total to £50k) (see Chap. 7).

District/unitary government policy gaps allow some room for innovation. There was some glimmer of bottom-up influence on degrees of niche–regime alignment. The LPA was perceived as the level that was not delivering on national priorities for self-build, which NDPs could circumvent with localised policies for delivery (although this is of course couched in central government policy being a key driver of self-build policies in the first instance). A local authority respondent suggested that innovation at a grassroots level becomes adopted in turn by the LPA (eventually) when emerging novel ideas can influence future iterations of planning policies, such as self-build and community housing—suggesting the content of bottom-up NDPs was influencing adjustments in the Local Plan (e.g., during Local Plan Reviews when LPAs are required to check their Local Plan is still up-to-date).[2] Though, when probed about the influence of the NDP on the Local Plan, one local authority officer felt that it was less about the values implicit in the policy than the pressure

[2] The NPPF sets out a requirement that policies in local plans and spatial development strategies should be reviewed to determine if they need updating at least once every 5 years, and then amended as necessary. See https://www.gov.uk/guidance/plan-making.

from the national government to think of alternative ways to deliver more housing, again highlighting the strong top-down influence of policy onto the scope and setting of NDP groups.

Consultants enrolled to support community-based plans may not always align on 'local' values and priorities or expectations on role delivery. Frome's Neighbourhood Plan enlisted various consultants to support its development (see Fig. 7.4). Yet governance arrangements with consultants at times were considered to engender broken links in the governance and overall process of the plan and there were some experiences of unidirectional and uncooperative relationships within the plan's governance. One local authority respondent suggested there were "power struggles" within the steering group and FTC over "who was in charge" and there was a tussle between FTC wanting the steering group to be stronger in challenging MDC while its members faced resistance to confronting planning policy and regime parameters. In part, there was a perception that consultants had encouraged unrealistic policies to be put forward. Indeed, paid consultants endorsed some more contentious ideas and gave FTC the confidence to supersede MDC advice, to "push forward" and ignore the discourse of "impractical" policies, for example, employment substitution policies, One Planet Living (OPL) principles and more ambitious (relocalised) low-carbon policies (see Chap. 7). The remaining disagreements between FTC and MDC were steered towards the plan's Examination as a space for mediation of contested claims in the NDP between the two organisations (as well as planning advisors) (for an excellent discussion on the role of the NDP Examiner, see Parker and Street, 2015).

Some more 'mainstream' planning consultants did not appear to be natural supporters of relocalisation, with one consultant not recognising the word "cohousing" after being asked several times in an interview with me about their thoughts on cohousing during the NDP consultation, while another referred to OPL in dismissive terms as "at this one world, erm, or concept or whatever you call it, concept, plan, way to live" (a quote that is illustrative in highlighting the vagueness of these principles in concrete planning terms). Moreover, another consultancy firm was felt to not represent Small and Medium-sized Enterprises (SMEs) in the town who had "business heads on" and focused instead on "the big boys" of employment; these particular consultants were felt by an IfF councillor to have overlooked the needs of SMEs in Frome and issues such as quality of space or location for different employment types, notably for the "small, homespun industries" (i.e., the creative, artisan businesses felt important to retain Frome's cultural identity). Some consultants spoke about how the "shoe was on the other foot" where they may have different types of clients simultaneously experiencing planning on "the other side"; that is, seeking ways to attain large-scale development gains for volume house-builders at the same time as supporting NDP groups like Frome's to encourage community-led development approaches. While these multiple roles were seen to assist knowledge transfer between roles (and different planning arenas) at the same time there was a clear value conflict where mainstream 'regime' client objectives contradicted the principles of relocalisation being sought within Frome's NDP.

There were also experiences of friendships breaking down, due to a perceived disloyalty within the performance of their role or seeking to add too much personal agenda into the process. One individual was felt to have overstepped the mark of an advisory role where their personal agenda compromised FTC's ownership of the plan as a result of their contributions by "railroading" their own views onto the plan. While, on the other hand, one consultant felt their task-specific Terms of Reference (ToR) didn't bring out the most effective contribution from their own experience and capacity and felt pigeon-holed in their role (see Chap. 7 for more discussion on roles in Frome's NDP). There was even an underlying expectation that housekeeping/domestic duties were reserved for females, leading to less "kudos" being bestowed on them.

The wheels of Neighbourhood Plans can spin without procedural alignment during their compilation, leading to later conflict or disappointment. As illustrated in Frome's case, often there can be high levels of enthusiasm of NDP groups to act with outside assistance (i.e., beyond their local authority via an NDP 'Link Officer' and using their own preferred consultants). However, this can disrupt—and bypass—potential linkages with the local planning authority, overlooking what one local official described as the "actual mechanics" of the planning process. A local authority representative suggested they could help to ensure early compliance with the process on procedural issues but there was a tension between "getting on with things" and "getting things right"[3]; they lamented that

[3]At the time of the interview, MDC was producing a support package to guide NDP groups to ensure links between the LPA and NDP groups wre made earlier in the process to ensure they get the right support to craft more robust objectives and help decide on whether an NDP is right for their community (e.g., whether a parish plan or village design statement may be sufficient in villages, or if an NDO may be more appropriate).

these missing links could have been mediated with more regular face-to-face contact between the LPA and the NDP group, or having a third party to review contentious issues between FTC and MDC (e.g., carrying out 'a health check' for policy compliance prior to submission). The lack of some fundamental procedural nuts and bolts, led one local authority representative to suggest that "there's something slightly wrong in the system in terms of the checks and balances there" in that "there's nothing to stop those wheels turning in the process". Again, this contrasts with Frome's Community Plan where there were synchronised and shared financial and governance arrangements between MDC's own strategy policy development and the Community Plan to support a co-dependence in the preparation of both plans.

Link Officers, as the gatekeepers of relational ties, can prevent structural changes of power dynamics and understanding on place-making issues between local government, local councils and communities. In addition to there being less contact with the MDC Link Officer than would have perhaps been desirable to expedite alignment between the town and district council on Frome's NDP, there is also a question of whether the role of a 'Link Officer' potentially undercuts the evolution of relational ties between tiers of government on local planning matters. For instance, the NDP process didn't encourage lines of interaction between FTC and more senior positions with MDC as these relationships were managed through the Link Officer (which is a standard governance arrangement for Neighbourhood Plan, see Box 3.1—The steps of the Neighbourhood Planning process). The actual involvement of MDC elected members (i.e. councillors) was considered to be "light touch". The planning policy team had delegated powers to engage with FTC to make decisions on behalf of MDC to expedite the process rather than being committee-driven (committees were generally perceived to be a highly politicised arena and were criticised by some Frome-based actors as cultivating party-political divides and decision-making).

More broadly, there was an everyday disconnection between the LPA and wider environmental groups—one officer had no contact with environmental groups since starting their role; another suggested that such engagement was directed (only) at the time of consultation on the Local Plan—thus missing the potential for continued constructive engagement with such groups throughout the planning policy lifecycle. The *Transition* movement was not apparent to one officer who saw 'transition' as belonging to the realm of development management rather than being aware of the broader social (environmental) movement. There was also doubt and hesitation about roles to deliver sustainability objectives within government. For example, who should deliver it and if sustainability is classified in planning terms beyond delivering green infrastructure (e.g., green spaces for nature or recreation) - discussions which could take on a more localised understanding if community groups were more aligned to planning policy; i.e., what key terms such as sustainable development really mean when applied to specific (and diverse) communities and how this overlays with their placemaking networks (i.e. natural, cultural and historic values).

Innovation often stems from path-dependent placemaking trajectories. FTC had sought to actualise the objectives in Frome's NDP through several initiatives, which in part built upon existing development agendas that existed prior to IfF taking power (such as pedestrianising Boyles Cross[4])—indicating how FTC's development and placemaking agenda represented a combination of both steer and path-dependent agendas. (This is a well-known phenomenon in the study of sustainability transitions literature and there is a wealth of studies on this topic. See for instance Hanger-Kopp et al., 2022 who examine the role of Adaptation pathway approaches (APAs) that can help to reveal and anticipate technological and institutional path dependencies).

Community Assets as a Seat at the Table in Placemaking: Influencing the Economics of Development and Investment in Landholdings

To have an impact on place, communities need to own or have a long-term interest in physical assets, for example, land and buildings. One New Economy respondent suggested that ownership was increasingly fundamental to realising *Transition* (relocalisation) - something that is increasingly enabled through the rights in the Localism Act (defined in sections 87 to 108 which give certain defined local bodies or groups the right to bid for a community building registered as an asset of community value and supplemented by The Assets of Community Value (England)

[4] See https://www.discoverfrome.co.uk/the-boyle-cross/ for historical photos of Boyles Cross (last accessed 26.1.19).

Regulations 2012). Taking an active, leadership role in securing financial mechanisms and control over physical assets in Frome was a key part of the town council's strategy to deliver on the objectives stemming from the Community Plan and influencing the tone of (major) development in the town.

Such a strategy also had the support of the business community who, according to one respondent, recognised the value of "old" and "new" businesses to mutually support one another (indicating both 'steer' and a mutual alignment on economic development)—again, terminology that echoes the distinction between the perceived social and political demographics of 'Old' and 'New' Frome (though individual local businesses were not interviewed to test this claim in terms of how they sat within such clear-cut distinctions).

Examples of asset acquisition in Frome by the town council or other active community bodies included the purchase of a parcel of land along the River Frome; this led to a partnership between Edventure and another local organisation to reskill unemployed people by building a shelter and improving a footpath, which could also be used by residents of a nearby care home and the wider community. Thus, providing a community-led approach which enhanced well-being benefits for multiple local stakeholders.

However, communities have an inherent resource imbalance when confronted with cash-rich developers, which means that many relocalised intentions failed to get off the ground because developers have the upper hand in terms of having resources to quickly put forward proposals:

> *[W]e can't compete with erm, multi-national housing developers on a site. They have instant access to wads of cash basically. [Tuts] … they have their own people looking for sites … They've pretty much got it sewn up. We don't really stand a chance … we don't have access to that kind of funding—for starters—so we can't just say [slams hand on the table once] 'Okay we will buy it.' (Local group representative)*

Saxonvale: A Beacon of Hope for Relocalised Development in Frome

Saxonvale has been a derelict elephant in the room in terms of Frome's politics and debates on local development. Those who sought to influence the outcome on this site were effectively vying for the future of the town's identity (i.e., can relocalised ambitions win over entrenched mainstream development resources and logics?). The idea of an 'alternative', community-led scheme for Saxonvale predated IfF and 'localism' policy artefacts, stemming from activity and contestation around the site dating back to the 2000s. A group of active (relocalised) placemakers also established the Carley Development Trust in 2013, in an attempt to seek "to bring together a credible development team behind a different vision for the Saxonvale town centre site in Frome". In an effort to influence perceived 'poor quality' planning applications for Saxonvale (notably by FTC and the group 'Keep Frome Local' that campaigned against a supermarket on the site), in 2013, FTC purchased a 0.57 ha parcel of land on the site (on Guy Fawkes night, deemed appropriate by one IfF councillor for their subversive ambitions) to protect the Somerset Skills and Learning Centre. This acquisition gave the town council a seat at the table for any negotiations on the site's future and to earn an income from any revenue arising from development on the site. This more asset-focused approach of relocalised development effectively led to "by-passing" the district council, as one IfF councillor suggested.

As described above, with a vested interest in the land holding (see the white lined section in Fig 8.1), FTC also commissioned the consultant to develop the Implementation Prospectus to find a way forward to resolve key issues in landownership and barriers to community-led development of the site which gave a focal point to discussions on the future of the site.

Underpinning these decisions was a pre-election visit to Frome by the Secretary of State for Communities and Local Government,[5] Eric Pickles, in May 2015 (owing to the fact that Frome and Somerton was a marginal Conservative seat) which gave the town council an opportunity to engage national-level actors and share ideas for alternative delivery mechanisms on Saxonvale. At this meeting, Pickles declared Frome as the "home of localism" (Harris, 2015). Pickles subsequently encouraged FTC to apply for central government funding which, as one consultant suggested, has "given [FTC] the confidence to sort of push on".

It is important to note that the context of the Saxonvale issue altered significantly when a change in the wider economic landscape led to

[5]The title subsequently changed to the Secretary of State for Housing, Communities and Local Government following the change of name to the Ministry of Housing, Communities and Local Government (MHCLG) in January 2018. In September 2021, the name changed again to the Department for Levelling Up, Housing and Communities (DLUHC).

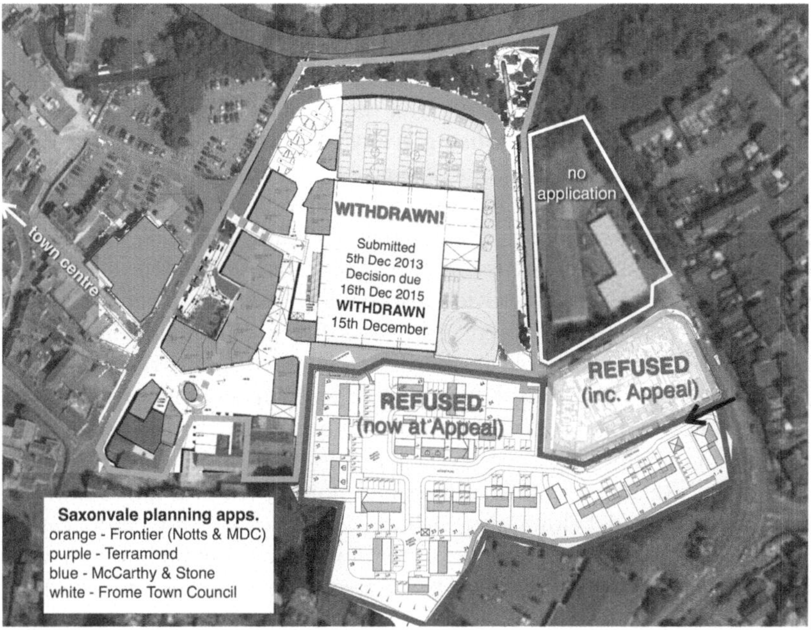

Fig. 8.1 Complex Landholdings on Saxonvale (as of November 2018, prior to the sale of the FTC-owned land) (Duke, 2018)

opportunities to re-evaluate landownership of the site. In 2018, the district council bought out all the existing landowners of Saxonvale with the aim of regenerating the site (until that point the site had been locked in a stalemate where three different schemes for housing and employment failed to get approved or built out). This included the three adjacent plots that were owned by Notts Industries (who had recently gone into administration), Terramond and FTC. FTC went on to sell their parcel of land for £985,000 (the sum of which was used in part to pay back a purchase loan and invest in Somerset Skills and Learning, a community support organisation). This meant that a comprehensive development scheme could come forward on the site (a much more cost-effective option than for FTC to seek to develop a relatively smaller proportion of land and the associated costs of commissioning architects and other consultants to design a scheme and put it through the planning process). An article in the Frome Times in January 2019 described how the town and district council were now "working together with the aim of facilitating the regeneration of Saxonvale". This purchase also led to negotiations with a property developer (Acorn Property Group) to bring forward a scheme for the site which ignited, once again, a battle over the site's future (discussed below).

The role of affordable housing is key to a just transitions model of development. Fair Housing for Frome (a registered Community Land Trust, CLT), established by IfF councillors and other members of Frome's active community to address the shortfall of quality affordable housing in the town, was also in discussion with MDC housing offers to explore whether the CLT could provide affordable housing (in perpetuity for 'local' people) on the proposed Saxonvale scheme. (Indeed the idea that a CLT would play a key role in the development of Saxonvale had been previously been put forward around 2015, which was supported by the HCA). Since then a Frome Area Community Land Trust has been established in an attempt to secure greater provision of affordable housing in the town. Related to more 'affordable' homes, the Mayday Saxonvale scheme (discussed below) also includes 27 detached and bespoke tiny homes, includeing 3.5m x 7.5m 'starter plots' as custom build opportunities for home builders that would be connected to a Tiny Home 'street' in the 'Tiny Homes Quarter'.[6]

Comprehensive doesn't mean relocalised development (if orchestrated by 'outsiders'). The change of asset holdings clearly paved the way for improved relations and the opening of communication channels on development between the town and the district council and led to the scheme that was developed by MDC and the Acorn Property Group. The Acorn scheme was approved in August 2022 by MDP's Planning Board (ref. 2019/1180/OTS), but this was still not to the satisfaction of the many of the relocalised networks and actors in the town. A very unusual situation arose where the Acorn scheme was granted planning permission in the same month as a rival, alternative scheme led by some of Frome's active placemaking networks.

A key issue with the proposed Acorn scheme was that it undermined the role of the Silk Mill, an undesignated heritage asset that hosts a creative and events space on the Saxonvale site. The Silk Mill was purchased in 2005 by the current owners from Notts Industries. The owners argued

[6] More about the Mayday Saxonvale 'Tiny Homes Quarter' proposals can be found at https://maydaysaxonvale.co.uk/masterplan/tiny-homes-quarter/.

that the Acorn scheme would effectively restrict access to the Silk Mill and its business severely impacted. The Silk Mill had largely been absent from any formal discussions with the different landowners of the Saxonvale site about its future development; at one point the owners had to fend off decontamination and demolition works associated with one proposal for the Saxonvale site that would have effectively flattened the Silk Mill, funded by a government grant.

To address this and other perceived issues with the Acorn scheme, a *Mayday Saxonvale* scheme - described by one of its proponents as "a Frome solution for a Frome problem" was developed, resourced and championed by the owners of the Silk Mill (Damon and Kate Moore) in partnership with others who shared their concerns (including a team of progressive developers, designers, landscape architects and financiers). Rather than 'business as usual' development and the crowding out of local cultural capital and community views on what was seen to be the 'right' kind of development for Frome—as is typical of all too many large-scale developments—the Mayday scheme aims to be an innovative joint venture with a developer that centre stages social value in their not-for-profit investment model for the site. Reflecting models in the US, the Mayday scheme seeks to reinvest profits into the local community in order to "fundamentally alter the viability equation". By offering a lower cost of equity and development funding, it is hoped the scheme can overcome the challenges raised by the local group representative about not having sufficient funding to own the assets required to unlock community-led development (for instance by using private risk capital and working with activist architects to make the scheme viable). Notably, the Mayday scheme seeks to cultivate "well-being generated by design" and "learning developed by construction"[7] and aims to improve local skills and employment in the implementation of the scheme (Fig. 8.2). The proponents of this scheme hope that if the scheme is successful it might be able to act as an investor of other community-led regeneration models, such as helping to fund community groups to pay for planning fees and helping to resource communities to grasp similar opportunities for alternative development schemes - as one respondent suggested, acting to "enlighten and enliven" other communities.

[7] See Tom Spencer's view on utilising community heritage assets on the site https://www.youtube.com/ as well as the overall scheme website https://maydaysaxonvale.co.uk/

Fig. 8.2 Core values and practices generating social value through the Mayday Saxonvale scheme (Mayday Saxonvale website, 2022)

The Mayday scheme managed to gain sufficient support from the community to progress this alternative scheme through the planning process—and had the support of FTC, Frome & District Chamber of Commerce and over 1300 residents—despite the resources invested by MDC to develop the Acorn scheme (which, in the view of someone close to the Mayday scheme was an example of the LPA effectively funding a private developer to undermine its own planning policies on affordable housing and employment space provision). In an effort of local demonstration against the concurrent Acorn scheme and a prefiguring (enactment) of the demonstrative and reciprocal character common to many New Economy initiatives, in May 2022 the Mayday Saxonvale campaign organised a mass 'hug' around the Saxonvale area. This involved around 400 people (stretching from the lower end of the site by Merchants' Barton car park, up and around Vicarage Street, and down to the bottom of Garsdale) as a peaceful protest in support of the Mayday alternative scheme. At a substantial cost to the Mayday Saxonvale campaigners, and after months of hard campaigning and liaison with planners, the Mayday Saxonvale scheme was also approved by MDC's Planning Board in August 2022. However, at this time the rival developer Acorn still had a conditional contract to develop the site and was finalising their plans for the riverside of the site (which is on a flood plain so can't be built upon) and negotiating with the new unitary local authority, Somerset Council who now owned the Saxonvale site, following the dissolution of MDC in April 2023 (for more on the changing administrative arrangements in local government in Somerset see Chap. 11). The Mayday scheme organisers launched a judicial review to challenge Acorn's proposals, which the Mayday campaign team felt would contravene the LPA's own planning policies and did not provide sufficient employment uses. The Mayday scheme won its judicial review challenge in October 2023 when the judge ruled that Acorn did not adequately address and represent the provision of employment space at the time their scheme was approved. This effectively quashed Acorn's planning permission for the Saxonvale site, which one person close to the scheme described as a "flat out victory [which] don't come along very often" where the scheme met its "David and Goliath moment". With Somerset Council facing critical budgeting issues the Council is looking to sell the Saxonvale site, leaving it open to Mayday (and potentially Acorn, if it continues to push ahead with its proposals) to purchase the site. Proponents of the Mayday scheme are exploring how they can resource the purchase of the site to make their scheme happen, including using social impact funding. A decision on who will win the prize of ownership on the site is expected in the Summer of 2024, one that will seal the fate of the 'existential' battle over what development model Saxonvale might follow.

The Saxonvale case shows just how resource-intensive using the planning system to challenge mainstream developer interests is and how to do so invariably takes highly committed, resource-endowed local actors to dream and continue to challenge the planning system—whatever this takes. The absence of Frome's Neighbourhood Plan in the legal case to prepare for such a legal challenge does, however, raise questions about how the planned revision of the NDP can help to strengthen the underpinning policy architecture that can bring community-led schemes to bear.

Conclusion

As we have seen, the planning process is a mechanism (as a tool or arena) to promote niche resistance to 'mainstream' development models. However, it also squeezed out (immediate) relocalised ambition and communities were not able to "change the thresholds" of the underpinning decision-making of planners or the profit-driven investment logics of most developers and landowners (such as pension funds) that often crowded out models for a lower return, impact investment model. Localism, while helping to stimulate ambition for community-led change and development did not necessarily change the underlying decision-making of planners and local authority officials. Some of the rights underpinned by the Localism Act - such as NDOs and CRtBOs - were not straightforward to use. Therefore, vested political and developer interests were favoured within an institutional arrangement that had several structural gaps and disconnected links between the local level and higher up the planning and political regimes.

Meanwhile, the administrative arrangements between tiers of government were perceived to obfuscate information flows and alignment of capacity and energy for relocalised placemaking. There was a clear sense that relational blockages needed to be moved on between government (such as historical tensions dating back to the time of regional development agencies). FTC was perceived to have steered the mechanisms for relocalised placemaking by securing financial stakes to influence relationships with developers and MDC (in the purchase of the parcel of land depicted in Fig. 8.1) and thus shifted the parameters of engagement by bringing the district council (and implicitly the other landholders on the Saxonvale site) *into* FTC's relocalised placemaking agenda (and it is important to note that alternative, community-led visions for Saxonvale existed before IfF took control of the town council in 2011). There was evidence that the influence of Frome-based intermediaries was moving on historical tensions from a competitive to a collaborative way of working which eventually led to MDC purchasing all parcels of land to enable a comprehensive scheme for Saxonvale. However, there continued to be a disconnect between the potential for innovation in what could be delivered on the site and the ability for planners to themselves support alternative development models, rather than more 'mainstream' responses to 'comprehensive development'. (In particular, the need to home in on reflecting the "DNA" of Frome—see Chap. 7—and its green creative and social capital to dream and enact place-based New Economy initiatives; though, as I raised earlier the term 'DNA' is highly contested given the different (classist) debates around "New Frome" and "Old Frome").

Due to the perfect blend of government intervention by the district to advance comprehensive development, having the right team to progress an imaginative scheme and progress it through the judicial review process, community support and a lot of hard work and dedication, the Mayday Saxonvale story—while still not yet built out and a 'done deal'—holds the hope and promise that the ideas of localism can enable communities to pursue development that positively reflects a creative, sustainable and dynamic future for communities such as Frome. But it's not just relational ties between tiers of government that need to shift. The Saxonvale case has revealed the highly exclusionary systems of land ownership and the politics of negotiating development trajectories with landowners that are largely motivated by (private) profit-driven development, which can undermine local alternatives of community-led schemes. The Mayday Saxonvale scheme offers a creative and innovative community-based impact investment model for Frome that can also inspire others. However, replicating these particular transition pathways is not always readily transferable in all contexts. If the Mayday Saxonvale scheme can deliver a successful blueprint of alternative, community-led investment then it might be able to mentor other communities to do the same. In the next chapter, we take a closer look at the role of informal ties within Frome's placemaking networks and how this influenced the politics of engagement in transition arenas, or disconnection with them and its implications for placemaking outcomes. This can shed light on the conditions that are necessary to enable an active cluster of community change agents to stimulate development alternatives and the challenges of bringing the wider 'community' with them.

References

DLUHC. (2023). Reforms to national planning policy report: Government response. 29th November 2023. Department for Levelling Up, Housing and Communities. Retrieved 6/12/23 from: https://www.gov.uk/government/publications/reforms-to-national-planning-policy-report-government-response/reforms-to-national-planning-policy-report-government-response

Duke, K. (2018). Landholdings on Saxonvale. Retrieved June 10, 2019, from https://saxonvalesite.wordpress.com/

Gallent, N., & Robinson, S. (2012). *Neighbourhood planning: Communities, networks and governance*. Policy Press.

Hanger-Kopp, S., Thaler, T., Seebauer, S., Schinko, T., & Clar, C. (2022). Defining and operationalizing path dependency for the development and monitoring of adaptation pathways. *Global Environmental Change, 72*, 102425. https://doi.org/10.1016/j.gloenvcha.2021.102425

Harris, J. (2015). How Flatpack Democracy beat the old parties in the People's Republic of Frome. *The Guardian*. London, Guardian. https://www.theguardian.com/politics/2015/may/22/flatpack-democracy-peoples-republic-of-frome

Parker, G., & Street, E. (2015). Planning at the neighbourhood scale: Localism dialogic politics and the modulation of community action Environment and Planning C: *Government and Policy 33*(4), 794–810. https://doi.org/10.1068/c1363

Parker, G., et al. (2017). Contestation and conservatism in neighbourhood planning in England: Reconciling agonism and collaboration? *Planning Theory and Practice, 18*(3), 446–465.

Wills, J. (2016). Emerging geographies of English localism: The case of neighbourhood planning. *Political Geography, 53*, 43–53.

CHAPTER 9

Informality and Engagement in Community-Led Planning, Placemaking and Politics

Informality and the Politics of Representation

Mediators and intermediaries—"Getting Frome" as a conduit for positive relationships. As we saw in the previous chapter, having an effective interlocutor that could champion Frome-led alternatives was a key enabling factor in unlocking complex development challenges. The extent to which individuals were able to *understand* Frome (i.e. as was mentioned by various Frome-based actors, its perceived immutable *alternative* 'DNA') played a vital role as a conduit for positive relationships in Frome's Neighbourhood Development Plan (NDP)—and beyond. A former Mendip District Council (MDC) Planning Policy Officer was considered to have a positive connection with Frome Town Council (FTC), partly because they were based in and grew up in Frome and "got" and "understood" what the town was about.[1] This was deemed a quality relationship whereby officers at FTC could "pick up the phone" to contact MDC. There was a perceived rupture in these relationships where turnaround in staff at the district council affected relational ties when this MDC planner left nine months into Frome's NDP process, which was considered to be a loss

[1] This individual also played a key role in the 2005 Saxonvale masterplan and was involved in other initiatives that cultivated strong, shared ties between MDC and FTC and the Frome community, including the Community Plan (Vision4Frome).

A. Burnett, *The Politics of Transition*, Palgrave Studies in Sub-National Governance,
https://doi.org/10.1007/978-3-031-40769-7_9

among Frome-based respondents and led to feelings of being "very much on our own" in the process and needing to "start from scratch" with a new MDC role holder (although the degree of substitution of support by their successor was not universally perceived by those I interviewed; some steering group members believed the new role holder was more or less like-for-like replacement and others felt they barely engaged with Frome's NDP at all due to the pressures of needing to prepare the district's local plan during the same period).

Crucially, there was also a sense that the facilitation of key processes during the preparation of Frome's NDP (notably the consultation process) ought to have been carried out by a Frome-based individual, rather than an externally contracted consultant; one facilitator's lack of prior connection to Frome was seen as a constraint, emphasising how a more procedural, formal approach didn't resonate with Frome's 'DNA':

> *I think it was because [they were] going through a formalised way of doing things which meant the connectivity with people in Frome was not particularly significant. [They] didn't know much about Frome, if anything ... I don't think we were short of [Frome-based] people who could do it either.* (Local group representative)

Connections to Frome were also felt to affect levels of support to the town's green and progressive agenda. Being Frome-based was considered to help with bridging the broader political divisions between the town and other levels of government, for instance the MP David Warburton (a prospective candidate at the time of the interview) was described by a Sustainable Frome spin-off group as "very *Frome* minded" (see also Chap. 8). People outside of Frome were perceived as more resistant to placemaking initiatives, such the proposed Boyles Cross shared space scheme. Indeed, a national-level actor working with independent councillors suggested local roots were a key factor for the longevity of independent councillors, who tend to "hold their own" and that good independents are hard to "dislodge".

Informal ties combined with formal routes into Frome's Neighbourhood Plan as entry points for shared agendas on relocalisation. Within the NDP consultation channels, a sense of (informal) understanding cultivated within Frome's placemaking networks flowed into the (formal) structures of the NDP to influence both self-build and cohousing in the plan (see Chap. 7). For instance, Figure 9.1, which sets out the

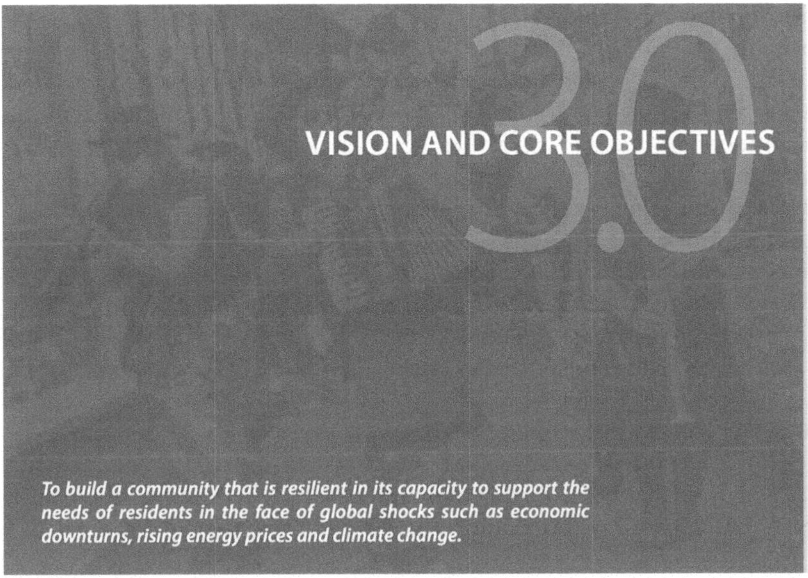

Fig. 9.1 NDP vision and core objectives (FTC, 2016, p. 9)

Frome's NDP's relocalised vision, is layered over a photo of the folk band *The Frome Street Bandits* (who describe themselves as "a semi-autonomous anarchic syndicalist community of musicians honking in Frome"[2] as a facet of the more radical expression of Frome's creative and 'autonomous' identity)—some of whom are actively involved in Frome's relocalised network (including Peter Macfadyen's wife,[3] Annabel, and the FTC Resilience Manager at the time).

There was a lack of tension, with seamless connections between New Economy actors which helped to forge clear shared agendas for what constituted appropriate development in the town. The similarity of values here was considered to feel like, in the words of one consultant, working in a town council "without opposition" given the extent to which some actors "chimed" or were "very much aligned in our thinking" on shared

[2] https://fromestreetbandits.co.uk/

[3] Peter Macfadyen established Sustainable Frome, co-found IfF and started the 'Flatpack Democracy' movement which sought to scale up Frome's independent political model. Like his wife, Annabel, he also has ties to the environmental movement Extinction Rebellion.

development philosophies to include in the plan and beyond, such as co-housing schemes and how these might be included in the plan. One consultant indicated that their own personal ties to Sustainable Frome led to an introduction to FTC and their eventual (formalised) contracted role as a consultant to the plan, since there was mutual alignment to use the NDP to instigate development alternatives. Moreover, an FTC staff member was also an active member of Sustainable Frome and had previous experience of working on One Planet Living (OPL), which later became a framework through which to explore shared Sustainable Frome—FTC objectives on sustainability issues (although, as we discussed in Chap. 7, OPL failed to be translated into Frome's NDP in policy-terms).

However, these informal ties were often conduits around the same, active people in Frome's placemaking networks—which raises questions about power and representation in local planning transition arenas. In recounting a debrief on the Citizens Panels—the main consultation arena to obtain feedback from the community in the preparation of Frome's NDP (see Chap. 7)—with another attendee, one respondent said:

> *[W]e probably said similar things and afterwards I might have said 'Well I'm not sure that went totally well because all the usual suspects were there.' And he said 'Yes, but that's the way life goes in Frome.* (Local group representative)

> *But actually, the reason those people are in the Org J*[4] *and erm, the Org AA and things like that [pause] is because they really care. So, they will come to meetings and if you try and get people who don't normally engage in that kind of thing to come along it's really difficult.* (IfF councillor)

Even if IfF weren't in power, active relocalised actors would have still taken the opportunity to influence the content of a Neighbourhood Plan or other community-based planning instrument. The influence of FTC (under the Independents for Frome independent group, IfF) on the engagement of the usual suspects—i.e. encouraging them to step forward into Frome's NDP consultation spaces—was deemed minimal since even if IfF had not taken forward NDP, these people would have likely been involved in trying to use localism rights, such as neighbourhood plans, such is their commitment to local placemaking issues. However, because of their commitment this meant that they tended to be more vocal than some 'ordinary' residents in Frome. For instance, one IfF councillor

[4] Org A and Org J are anonymised ids given to two particular Frome-based groups.

relayed an account that indicated the usual suspects might have "gate-crashed" the Citizens Panels and sought to control aspects of the process, which unsettled some (novice) participants.

Although one consultant external to Frome said they didn't recall that environmental groups were "over-represented" nor that as a consultant that they were given a steer to focus on environmental issues in the process, one local group representative thought otherwise, leading them to question whether there was a truly inclusive governance approach towards the 'non-green' perspectives in the town:

> *I'm aware that quite often you do get your usual suspects coming along to meetings and sometimes they can be a nuisance and sometimes they can, they can ratify everything that you want to do. And in this case, it just so works out that because there's lots of trendy, interesting people living here that are, er, as part of the Sustainable Frome network or Transition Network, that they are, they are applauding what, what the er, erm, what the council is doing which is great but I'm just a bit concerned about, as I keep saying, about them taking everybody else with them ... to me the way you do something is almost equally important as the end result, if you see what I'm saying.* (Sustainable Frome member)

Experiences of participation are varied and subjective in community-based planning, depending on personal preferences and positionality within local networks. The experience of community participation in Frome's NDP process varied for respondents, with some feeling that they were actively excluded from formally participating (for instance, given an opportunity to be part of the steering group), or on the other hand, in some cases some actors described how they gained frictionless access to the NDP's stages of development. Some community groups with previous experience and interest in planning were enrolled in the NDP in a public-facing capacity, with more informal discussions for those who wanted to keep a lower profile in the town. Any engagement with or through MDC was deemed by an IfF councillor as "dead boring" and not reaching beyond "anyone other than who you'd expect them to engage with" (compared to the more co-dependent consultation in Frome's Community Plan, described in Chap. 7).

Governance of restricted access can lead to charges of "incestuous" relational ties. However, the decision to keep the governance of Frome's NDP restricted led some to question the transparency and openness of the steering group. Two respondents keen to get involved in the NDP process were either purposely directed to input through the Citizens Panels or unable to attend them due to conflicting priorities, despite feeling that they had valuable skills and relevant experience that they could contribute to the plan. They respondents reported feeling “outside” or excluded from the process at both an organisational and a personal level, including one Sustainable Frome member. The overwhelming representation of FTC members or IfF councillors on the steering group (i.e. all members of the steering group except the Chair of the local Civic Society) led to a perception of secrecy, of (again) “incestuous” formal ties by one local group representative (a term also referred to in relation to Frome’s closeknit placemaking networks), a sense of unidirectional engagement and a lack of ownership of the community around the process. One respondent felt that self-build was being “pushed” onto the plan and through the Citizens Panel they attended, and indeed one consultant indicated they themselves were also “pushing” self-build onto their neighbourhood planning clients in general, due to its increasing prominence on the planning policy scene. This frustration is reflected in a local group representative’s comment that:

> *Then the town council had no interest to me being party to [the NDP]. Which seemed a bit odd … ’We’ve got to keep the group small and effective, so we don’t want lots of other people. So, you can have your chance in the Citizens Panels’ … it had nothing to do with [party] politics. I think it was they had a particular idea of what [they wanted to be] was in it … I just think they wanted to do their own thing. Because Org J, the person or the people involved, were very much involved in setting up IfF as well. So, it was a little bit incestuous. They presumably wanted to put their character on what was going on.* (Local group representative)

There was also a sense that consultation of the ‘usual suspects’—purposefully or not—ratified the personal agendas of IfF councillors without a broader endorsement of the wider community (despite FTC’s best efforts to encourage people to attend the Citizens Panels), a process that became in the words of a Sustainable Frome member “self-referential”. An event on self-build and community-led housing at Rook Lane Chapel in September 2013 featured the well-known designer and television presenter Kevin McCloud to corral enthusiasm on self-build (see Fig. 7.4) and was linked to a strong pre-existing commitment to pursue self-build development models by some steering group members.

Innovative governance methodologies need careful facilitation. Despite being heralded as a success by both the Examiner’s report and MDC respondents, the Citizens Panels still had modest attendance (with attendance varying between 12 and 30 people in each one) in the four panels (see Bourne, 2012 and Chap. 7) and were seen by some I interviewed (including an IfF councillor and a consultant) to be unrepresentative of wider social groups in the town owing to relatively low turnout, despite using methods such as randomised leaflet drops in key areas.

However, there was a mismatch between the extent to which the Frome's Neighbourhood Plan consultation process reflected the more progressive approach of FTC under IfF, with one respondent suggesting that the process became "too ordinary", "lacklustre" or operated in "a very standard way" (suggested by an IfF councillor, an FTC representative and Sustainable Frome member), despite World Café (a governance technology of *Transition*) being used in the Citizens Panels. Or in the words of one respondent who attended the Citizens Panels they were "jolly" and "fun evenings sticking bits of paper everywhere". Moreover, the Citizens Panels were felt to be too prescriptive (exploring the relevance of the Community Plan's priorities)[5] and not a more bottom-up, open starting point—thus the experience of participating in the Plan felt very different depending on who you spoke to. For instance, one respondent felt that the Citizens Panels were too premature and lacked a vision, objectives and draft policies to allow for a richer debate while others recognised that the Community Plan was a good foundation to start from (and this was of course IfF's 'de facto' manifesto commitment and from a political point of view, a good way to blend community-led planning and political innovation through the arena of the new opportunity to develop a neighbourhood plan.

Sustainable Frome Contributions to Frome's NDP

The procedural delivery of Frome's Neighbourhood Plan affected some local innovators motivation to become actively engaged. In the earlier stages of Frome's NDP's development, in an effort to drum up support within Sustainable Frome to engage in the plan as an organisation (in a similar way it had done with the Local Plan), a director requested a representative from FTC speak at a Sustainable Frome meeting. However, rather than making a formal connection between Sustainable Frome and the NDP process, this meeting failed to engage members of the environmental group; it was perceived to be "abstract", dry and lacking connection to the organisational objectives of Sustainable Frome—on this occasion Sustainable Frome and FTC were *not* on the same wavelength. This may have been due to a misalignment between the messenger and the receiver, where the new

[5] Which was the consultant's brief.

role occupier in the town council was not felt to have the organisational objectives of IfF being, in the words of Sustainable Frome member, sufficiently "seeped into" them and therefore may have come across as possibly less progressive in their delivery of the potential of the plan to connect to wider (relocalised) environmental agendas. (Here, a Sustainable Frome member wondered whether this FTC-Sustainable Frome interaction was more of a box-ticking exercise, rather than genuine alignment to garner Sustainable Frome's support in the plan—a similar charge that was made by IfF councillors against MDC and higher tier engagement with lower tiers if government). This disconnection led to a unidirectional engagement approach with individual members submitting their own personal comments as part of the formal Reg 14 process or through participation in the Citizens Panels, rather than an organisational engagement between FTC and Sustainable Frome (for instance as a formal steering group member or more regular structured interactions between them).

Former ties to Sustainable Frome were used to justify 'community engagement' in the Neighbourhood Plan. While Sustainable Frome's engagement in the plan was not "formalised", as a member of Sustainable Frome relayed, the name of Sustainable Frome still appears on the front cover of the Neighbourhood Plan.

This was considered to give false impressions of Sustainable Frome playing a specific role in the plan and raised doubts over degrees of consent and legitimacy through the prominent use of its name. In addition, the transparency of the process overall and the legitimacy of an organisation speaking on behalf of its members without full endorsement was questioned by a Sustainable Frome member:

> *So, so actually I think, I think overall it's, it's a great document and it's absolutely taking things in the right direction. Erm, and, but my issue is around how you take the community with you ... in terms of liaison, I was a bit concerned about how Sustainable Frome's name, not just appeared on here but appeared on other things, about how that is ... And I think sometimes it's taking liberties ... Thinking, oh well Sustainable Frome in principle would support this ... [and] just pull Sustainable Frome out of a hat sometimes.*[6]

[6] In response to this point when sharing my PhD thesis for comment, one individual close to the process suggested that it was partly because there were so many people involved in the plan's development akin to "writing a novel by committee"; at times the draft plan included some unapproved ideas on relocalised development that had to be re-aligned (i.e. in order to seek to meet the basic conditions and steering group input) to fully evidence and justify its content, as discussed in Chap. 7.

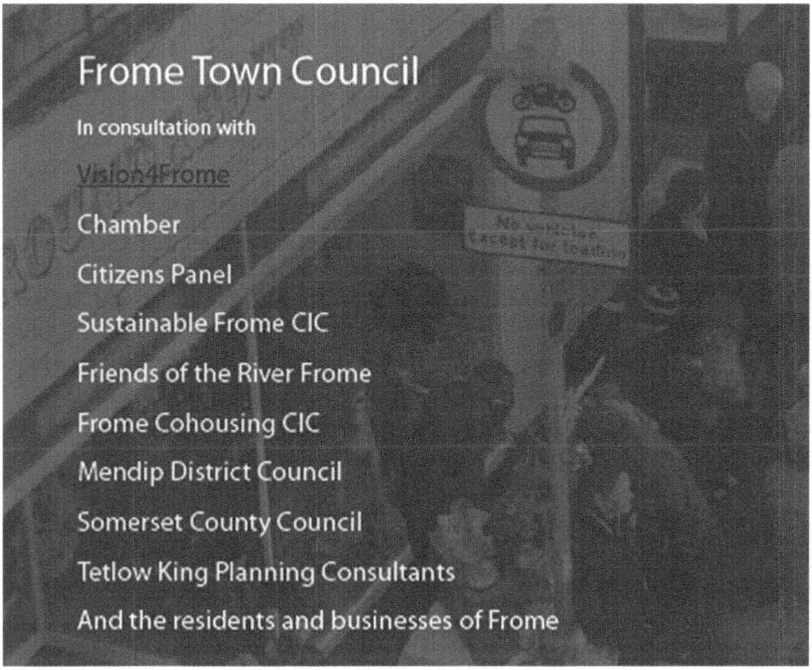

Fig. 9.2 NDP front cover with V4F added (FTC, 2014, p. 1)

It is noted that—despite the strong role that it played in acting as IfF's political agenda and structuring the development of the NDP, notably in the Citizens Panels—there was a complaint that Vision4Frome (reflecting the Community Plan's Management Group) had been absent from a list of consultees in previous versions of the plan and there was a request to include it, which was added with track changes in the September 2014 version of the plan (V2; see Fig. 9.2). (This is the opposite experience of using Sustainable Frome's name when the group felt they hadn't necessarily bought into the plan, which may suggest the politics of using, claiming and justifying relocalised ambition using community-led planning tools; while not intentional—where in the effort to do so previous (path dependent) events may not always get fully recognised).

Barriers to Inclusive and Representative Politics and Placemaking—Representation Below the Parapet: Citizen Interests, Class and Gender

Planning isn't 'exciting'. Planning was deemed by an FTC representative as not being "exciting" enough for the young or schools to engage with it (despite FTC having a Youth Mayor and Deputy Mayor, which one might assume would indicate higher than average engagement of the young in Frome's politics and placemaking). Therefore, it often fails to reach a wider audience to shape the inputs into planning, despite a wealth of participatory planning tools and frameworks (see Hickey and Mohan, 2005).

The NDP process did not sufficiently engage everyday interests because it wasn't felt to be 'fine grained' enough or was itself considered too 'niche'. As we have seen, despite making inroads in stirring interest in Frome's political approach the extent the broader community were willing to involve themselves in NDP consultations, including the Citizens Panels didn't really break new ground in terms of garnering interest of participation in planning. The FTC-led Reg 14 (Pre-Submission) consultation received 50 responses while the MDC-led Reg 16 (Submission) consultation received 16, mainly from statutory consultees (i.e., organisations that have a duty to respond) or who had some kind of skin in the planning game). Though, despite originally being concerned about whether or not consultation had been carried out sufficiently given the relatively low responses on the plan, one FTC respondent was satisfied it meant "there's not masses and masses of objections".

However, one steering group member suggested that the actual scope of Frome's NDP could have been a reason for the lack of interest by many people in the town as the plan was not fine-grained enough to cover a town of its size, "for a lot of the people of the town, it's got no relevance whatsoever". Thus, there was a trade-off between detail and strategy, even though Neighbourhood Plans are meant to offer greater local flavour for (some) planning policies. Wider feelings of disinterest in the Neighbourhood Plan were amplified when respondents spoke about their attempts to engage ethnic minority groups, such as the Polish community and the traveller community (where it can be challenging to win their trust due to their frequent mobility). Interestingly, an IfF councillor described planning as "quite niche" when they reflected on attendance at an 'open source' planning event at Rook Lane which also didn't get a large turnout, but a committed number of people attended. Many sustainability transitions scholars have tended to use this term to relate to more grassroots rather than 'regime' actors and arenas. Yet as this respondent's use of the term indicates, the framing of planning as a "niche" arena reflects that the politics of what is considered niche is relative and definitely not a binary 'niche' vs. 'regime' (mainstream) clear cut distinction, again highlighting the subjective qualities of the positionality of development alternatives.

People only show an interest in planning when it affects them (i.e. their interests or values). It was felt that was only when planning affects people personally will they start to pay attention, otherwise people (unintentionally) delegate their agency to influence placemaking and development issues to those who will stand up because they have interest and energy to—which a local group representative suggested leaves the usual suspects to have a "disproportionate impact" on local politics. For instance, a young planning apprentice that did show an interest in planning had personal ties to the NDP steering group chair, meaning that in some cases for a younger—and more in experienced—individual to step up in placemaking arenas requires an activating ingredient of social capital and mentorship. (Compare this to the way youth participation was utilised through social media channels and enabled the youth activist Greta Thunberg to inspire a generation of young people to engage in environmental action - known as the 'Greta effect' (Sabherwal et al., 2021) and spurred the youth climate protect movement, Fridays for Future (see Wallis and Roy, 2021). Thus demonstrating how social media can help to stimulate campaigning agency and the underlying conditions underpinning the potential for the younger generation to engage in issues that matter to them (Belotti et al., 2022).

However, linking these activating channels to placemaking and planning can still be 'blackboxed' issues (see Rydin, 2012) and there are obvious challenges to democracy by social media, as we discuss below. An IfF councillor questioned whether the fact that if people are so disinterested in planning—despite offering small financial incentives to attend a Citizens Panel—but others who do really care show up (i.e., the usual suspects) does it really "matters" in the end that they have more influence, because they are committed; as they suggested, "it's a cynical view but what else do you do? What other conclusions can you make?". As a Frome-based respondent close to the Mayday Saxonvale scheme suggested, "communities aren't naturally occurring, like apples on apple trees or raspberries on raspberry canes. They need a focus, something neighbours can identify with and participate in doing together". The experience so far of planning processes is that they are not yet sufficiently structured to allow a range of diverse opinions, capacities and interests to be translated into the mechanics of inclusive and representative transition arenas.

People don't always think enough into the future to care about the purpose of planning. In addition to the structural or administrative challenges to participation raised so far, the issue of low motivation correlating to a mismatch between the long-term, more strategic, aspects of urban planning and development and the more immediate concerns of the here and now (also known as the 'value-action gap' (see Franco and Ghisetta, 2022) was identified as an issue. This lacunae was felt to affect the extent people got behind the consultation process in Frome's Neighbourhood Plan. Lots of studies have been carried out on the issue of 'consideration of future consequences', particularly in relation to pro-environmental behaviours (see Arnocky et al., 2013, who found that less pressing immediate concerns can affect people's willingness to think more into the future). The general perceived lack of forward-thinking placemaking visions by the wider community—structured by the longer-term horizons of planning policy frameworks—is summarised in the words of an FTC respondent:

> *[Planning's] not an exciting topic ... a lot of people—I don't think they look to the future and realise that the policies that are put down now in 20 years will be affecting how the town grows ... [the young] just don't seem to look that far ahead. Or appreciate that the planning process can have an influence on how their town might develop in the future.*

Procedural timeframes can undermine the longer capacity building strategies to empower communities to implement their own development. At a more general level, there was a clear trade-off between deep engagement, as one New Economy actor suggested, to "agitate tensions", building local capacity to meet development solutions themselves and the speed with which to have anticipated development controls in place. There was a sense that an active, engaged and co-produced relationship with citizens was still far off from the majority of residents either getting involved or caring in the first place. For deeper community engagement, one

Sustainable Frome member suggested that people need to be informed and educated to participate fully, which requires commitment to participatory democracy—both by citizens and local governments. Indeed, the use of World Café governance in the Citizens Panels was felt to be inappropriate given the baseline capacity to engage in placemaking issues. (In World Café scenarios, the topics for discussion very much follow the lead of whoever in the room wants to propose a discussion based on suggested topics that other people then crowd around and self-select which group they want to be in). Such open governance was cited as one of the reasons that discussions weren't more targeted or more realistic and attracted more "quirky" ideas than practical solutions.

Some respondents pointed to a capacity-action gap (similar to the 'value-action gap' referred to above) or that the perceived influence of developers' ability to challenge planning decisions renders community aspirations meaningless in a David and Goliath scenario, where many might give up on thinking they have agency to change this dynamic. Frustration with 'consultation fatigue' (when planners, local government or other organisations constantly invite people's thoughts, and this doesn't necessary lead to any substantial change) and no workable delivery for the Saxonvale site in Frome for many years (see Chap. 8) led to a feeling of pointlessness of the planning process to deliver anything tangible—in its most extreme case, this led some to believe that there were pockets of a disengaged and despondent community in the town that the planning (consultation) process had failed to serve. At the same time, there was also a sense that asking people what they want gives impractical suggestions or prevents sounder ideas from gaining traction. As an FTC representative suggested regarding participation in Frome's Neighbourhood Planning consultation arenas, the Citizen Panels:

> *Yeah, I mean, this, if you do one of the exercises where you say to people how do you want Frome to develop in the next 20 years, you get animal parks, you get bowling alleys, ice-skating rinks, things that are completely actually impractical. Where they can go? Who's going to pay for them? And it's great that you want to see those things, that, that's the problem with asking people ideally what they'd like to see as opposed to what realistically can happen. (FTC representative)*

A deeper disconnection with politics and lack of effective information flows can cause distrust or disinterest with planning. Linked to the above issue of having sufficient capacity to engage in placemaking debates and planning consultation arenas effectively is the issue of what this has on the wider spaces for democratic debates and who gets heard. For instance, poor information flows that can lead to resistance to change or more 'progressive' outcomes. Talking about a failed experience to innovate by a council in a neighbouring county, one respondent suggested that democracy can get undermined by poor information flows:

> *Democracy is great and engagement is great as long as people are properly informed, and they properly understand the consequences of er, voting against something and they understand that erm, that you don't end up in a situation where you just kind of descend to the lowest common denominator, that actually people aren't prepared to listen to you … Particularly at the town council level there's quite poor information [flows] and councils are often quite bad at communicating with people … and then you get this kind of Chinese whispers thing … And then, you know, you start—there's this whole kind of, a wave of no, erm, you know that resistance to change, and [intake of breath], and, and yeah the consequence being that it doesn't happen at all.* (Local group representative)

Social media doesn't always help to amplify civic issues, such as community-led planning. The rise of the social media giant Facebook was generally felt to have a much more active reach than the town council could have through their own media and communications. However. an FTC respondent suggested social media is generating more "superficial communities"—giving rise to things that get people's attention, but generally these weren't on the topics of planning for the future, as one local group member suggested, generally being on quite "trivial" and "insular" matters (i.e., an abandoned campervan attracted greater interest on Facebook than the NDP[7]). (This is echoed in recent studies on the role of social media on democracy: Lorenz-Spreen et al., 2023 found that social media generally had a positive influence on opening debates in emerging democracies or autocracies but less so in established democracies, which indicated a rise in popularism or a weakening of trust in relation to social

[7] About 50 formal comments at Reg 14 compared to 15,000+ people who followed this issue on Facebook (which attracted hundreds of comments in 2015).

media use). In Frome's case and in the words of one town council representative:

> *And we're like, so neighbourhood planning like the next 20 years of how the future of Frome will develop, not interested. Somebody's abandoned their campervan, 15,000 people have read it .... They're not living in it, it's broken down, so you know, that's why they can't move it. But it's just why people have clung onto that story?* (Town council representative)

However, there was a recognition that intergenerational disparities (i.e., difference in perceptions, behaviours and capacities of younger and older people) may affect how social networks are built across the different demographics. However, while technology can enable self-employed people to work flexibly in continuous globalised connections, technology and modern life is a cause of information overload, where as one IfF councillor suggested, people are wrapped up in their own "little lives", which can hamper a broader collective understanding of engagement in the world and discourages people from engaging with planning:

> *We are all busy, we all have massive pulls on our time and thousands of people trying to grab our attention all day every day. Advertising or whatever it might be. We blank it out, we blank out everything that doesn't really concern us. And we just look at what is on our menu for today. I do anyway. And people have got other concerns and they think, what's the point?* (IfF councillor)

An FTC representative also reflected that social media platforms, noting one in particular, "had become a mechanism for protest" and complaint, which have a different flavour to targeted campaigns led by the *Keep Frome Local* (anti-supermarket) and the *Frome for All (pro-supermarket)* debates on Saxonvale in the 2000s. This suggests a splintering of traditional movement campaign strategies. However, the Mayday Saxonvale campaign (see Chap. 8) also highlights the critical role of social media and other platforms used by communities to galvanise interest and resources (such as crowdsourcing platforms) in corralling communities to support local development issues, and alternatives. However, mobilised community efforts for alternative development schemes via social media need to be supported by the underpinning planning and political structures that enable people to cooperate (someone close to the Mayday Saxonvale scheme lamented that "perhaps people are just unwilling to work together for the common good—something we are seeing writ large in UK politics at the moment"), calling on more attention required by policy makers on bridging interest and enabling conditions for communities to be truly empowered (something that might be further encouraged by Labour's 'Vision for Community Power') (see Comer-Schwartz et al., 2023).

Most people don't 'connect the dots' between active placeshapers and positive action in the town. There was a sense of surprise that 'ordinary' people are impressed by things that happen in places like Frome but have never considered the role of the town council or others involved in getting these kind of initiatives of the ground. A failure to reflect on how things come to be among residents (and visitors) in the town can cause a separation between valuing local councillors and their role in shaping cultural identities. As an IfF councillor lamented:

> *I suppose they kind of know stuff's going on in Frome, but [hesitates] the thing I find so odd is that people don't join things up. So, they'll come to the market and go whoa this is fantastic, 1000s of people in Frome, it's brilliant sort of thing. 'So, [has FTC] had anything to do with this?' 'Yeah, it wouldn't have happened except for us'. 'Really? God, that's amazing'. But they haven't thought that through before. There's not those connections made. I think [people] genuinely haven't thought how these things might have happened or why they might have happened. Or, things like that.* (IfF councillor)

Participation in planning can be tainted by classist persuasions. These factors were considered reasons why "Old Frome" lies dormant, and occasionally "wakes up" when they see physical changes in the town as a result of placemaking by the very active. While one local group member dismissed the perceived distinction between Old and New Frome—indicating that this term was a socially constructed divide—for some respondents, Frome was serving the interests of the middle-classes well, but there was a demographic divide between incomers "New Frome" (Down from Londons, DfLs or 'blow ins' following the COVID-19 pandemic and the rising cost-of-living crisis) and born and bred Frome "Old Frome" Speaking in 2015, one Sustainable Frome member commented:

> *And at the moment there seems to be an attitude in the, with the new IfFie councillors who—everybody thinks they're marvellous—all the middle-class intelligentsia think they're marvellous, right … it's attracting middle class people to live here … the IfFies have identified there is a poverty issue in Frome.*

> *Erm, what they don't ever talk about is class. Nobody likes to talk about class anymore. But, erm, I, I actually think there is a bit of, it is a division along class lines, frankly.* (Sustainable Frome member)

When I interviewed people in Frome, I also found some implicit classist assumptions about people's capacity, such as that the Sustainable Frome member implied that helping the 'working classes' is dependent upon the middle classes supporting them, suggestive that placemaking is a classist arena requiring education, skills and time/resources. One local group representative respondent suggested that the 'nanny state' was also considered to stymie society's collective potential (and mirroring the discourse at the time of the Cameron Conservative government), making people lazy and unambitious. Though this is obviously a heated issue that is the subject of other books than this one.

However, as we will discuss in the next chapter, IfF's politics is helping to transcend political and social identity barriers. For instance, in 2021 Andy Wrintmore—a drummer in a hardcore punk band SickOnes—became the Mayor of Frome. In an interview with the rock magazine, Kerrang (Hutchcraft, 2021), Andy described himself as "Old Frome" but able to appeal to both sides of the Old/New Frome divide. This article showed that on the one hand he appeals to New Frome (his rock dress attire and beard reflect some visual identity markers of 'DFLs' or the fact he was elected as a punk rock mayor appeals to why they moved to the town), he also appeals to Old Frome (he works in the local Co-op supermarket, he drinks down the local pub and he 'talks like them').

Yet there is a backlash against gentrification and its erosion of 'edgy' social and cultural capital. Amid a plethora of recent newspaper articles that claim Frome has most definitely gentrified some residents, including some of the more established New Frome incomers, are following others in places like Bristol that have seen continued gentrification and its adverse effects on place identities, declaring they want to see their towns and cities "made sh*t again"—an obvious pun on Trump's slogan "Make America Great Again" (see Håkansson, 2018 for a discussion on the politics of gentrification and sustainability transitions in Peckham, London). How 'Old Frome' and settled 'New Fromies' can reclaim their representation of Frome's place identity remains to be seen. (One established New Frome resident who moved to the town over twenty years ago said that each phase of incomers brings with them their 'ways' which seem different, suggesting that demographic and identity shifts are part and parcel of healthy change to place-based identities, see Kyle et al., 2014). However, the fact that IfF's politics clearly invites a meeting of both sides of the Old/New Frome demographic split is encouraging.

Planning policy actively shapes the demographics of place. Further elaborating issues of class and demographics, the drive for affordable housing under the strategic policies of the Local Plan was seen by a Sustainable Frome member as a disruptor to the broader trend of New Frome moving to the town, that is, affordable homes were seen to be for "lower middle-class people" (i.e. that affordable housing, when delivered, can be a mechanism to artificially put the brakes on gentrification processes). However, the provision of affordable housing continues to be undermined by arguments of the viability of schemes, despite the cost-of-living crisis and the price hike in rental costs—the UK homeless charity Shelter estimates that almost 2.5 million renters are struggling to pay their rent—up by 45% since April 2022 (Shelter, n.d). In Frome, the average rent has risen to £1,499 a month—50% of the average salary (see Frome Times, 2023). As a result, some councils have declared a 'housing emergency'. In Frome, FTC have declared a housing crisis and has pledged its support for the Frome Area Community Land Trust to provide for affordable housing through this model (i.e. a community-led planning mechanism). The town council also seeks to include provision of 40% affordable housing on development in its revised NDP (although the amended plan will still need to comply with national policy; notably paragraph 65 of the December 2023 National Planning Policy Framework (NPPF), which sets out conditions relating to affordable housing on smaller sites). More recently, the delivery mechanisms of lower-cost housing have been given an alternative flair in the town, with proposals for the UK's first tiny homes community proposed under the Mayday Saxonvale scheme (see Chap. 8)—though its is not clear whether these homes would cater to the needs of "Old Frome" and how this alternative housing mechanism will perpetuate or help to solve the complex issues of genuinely affordable and inclusive housing in the town, and beyond.

People don't always have the right channels to participate in planning or placemaking. There was a sense among those I interviewed that there were insufficient continuous *conduits* to respond to planning processes. For instance, the staged structure of Neighbourhood Plan and Local Plan consultations can discourage people to engage in more formalised channels, or if people are not continually engaged and their interest wanes. There were no perceived active mechanisms for 'working-class' people to be meaningfully engaged in placemaking or planning in a way that effectively included them as actors in wider placemaking networks, beyond the specific channels used by the 'usual suspects'. However, one respondent believed that with appropriate role models, a more active engagement in placemaking could be learnt to help others experiment with new ideas where they live. Another local group representative expressed a need for people to take greater responsibility for their actions and enact a duty to become more engaged in placemaking.

More recently, there has been a widening of placemaking engagement channels in the town—notably through the Frome Neighbourhood Network (FNN) which seeks to create a neighbourhood, street-by-street approach to fostering inclusive, resilient and caring communities. This is potentially a very effective model to address the lack of conduits to engage in placemaking identified when I carried out my original set of interviews in 2014–15. This group is also utilising social media as a means to blend physical and digital engagement spaces to encourage inclusive arenas of mutual support (WhatsApp hadn't hit the digital scene at the time of these interviews and has since come to play an obvious key role in stimulating conversations among previously unconnected groups). The FNN has set up various neighbourhood 'toolkits' to facilitate communities in the town to share more information and create common spaces and events to encourage more connection between them. This model is also greatly helped because FTC has resourced roles to support these 'grassroots' models to emerge; for instance, there is a Community Development Projects Officer at the town council (and again points to the strategic role that FTC plays in facilitating wider placemaking networks in the town). There are ties to the relocalised placemaking network and the *Transition* movement in the FNN steering group, via Annabel Macfadyen.

This neighbourhood-level structure also raises the possibility of a much greater level of engagement and connection to the planned update of Frome's Neighbourhood Plan, beyond the more arbitrary and temporary groupings in the four Citizens Panels to a much more inclusive and wholesale connection to the town at a genuine neighbourhood level. Time will tell how this unfolds within the constraints of the planning process already mentioned in this book. But whether a more neighbourhood level network can offer more effective connections between placemaking and planning spaces (such as the nascent Somerset Local Community Networks, LCNs that seek to bridge the governance gap between the new unitary and local administrative levels) is something that will reveal the extent to which community-led governance can better support planning for transitions.

Women's participation in politics is gendered. I found that (negative) gendered role assumptions were considered key reasons why women are not considered as 'experts' in placemaking and planning, which can be exclusionary for some women to access leadership roles in their communities. This contrasted with an experience of one female I interviewed who had previously worked in a European country, where she noted that respect for the role occupier in a profession helped protect women from gendered inequalities. However, in the UK she felt that negative gendered role assumptions were seen to deeply affect women's equality of opportunity and standing in a professional capacity.

The conditioning of past experience was also considered to affect women's entry into political roles, including fear of being judged, the sensitisation of family and gender roles during childhood, school and university as well as diversity within the workplace. This conditioning hampered some women's potential but it also opened up role conflict to allow for reflection to unlock former women's potential contribution to public and political life. Women were also considered to be put off from engaging in politics because, in the words of an IfF councillor, of a "very male-dominated … stuffy, public school debating society" style to UK politics, even at a local level; as a result, some women fear being "judged" if they put themselves forward (although this obviously applies to some men too). In 2011, seven men and three female IfF candidates were elected in the Frome Town Council elections,[8] despite a 50:50 balance between men and women as potential candidates. One respondent suggested this was due to 'bad luck' of the electoral ward profile where some women candidates stood (and lost). The election of more men than women in this election was felt to hamper diversity and creativity in society more generally as an implicit imbalance between male and female political role holders can become, in the words of an IfF councillor, a "self-perpetuating problem, and that's bad for democracy".

Some women were motivated to get involved in local politics to be a role model to their daughters and challenge sexism. Interestingly, one respondent suggested she was unable to perceive gender dynamics for young girls because she didn't have daughters of her own. There was a

[8] This compares to nine men and eight women who were elected in the 2015 election.

personal account of misogyny as a barrier to entry to engage in placemaking in one female respondent's own community (outside of Frome)—despite having valid professional skills and experience, gendered views were preventing her and other women from being 'seen' by some male parish councillors. While another respondent assumed that the Neighbourhood Plan Examiner was male (she wasn't).

Despite conscious efforts to take a more inclusive approach to women's participation in politics (particularly in IfF's second administration, 2015–2019) I also uncovered instances where women appeared to be unseen by men. For instance, when asked, one FTC Officer suggested that two male FTC staff and councillors were active in distributing the IfF model. Yet they failed to see the efforts of a female councillor in disseminating the model and her active enrolment of creative planning professionals and feminist actors into the fabric of Frome's local politics and planning arenas. Moreover, despite a commitment of IfF to diversity, autonomy, and inclusion there were instances where some staff or consultants experienced gender exclusion within the NDP process, as one female commented:

> *I did periodically feel that, erm, I wasn't given as much kudos in terms of what I actually did for work, as perhaps a male counterpart might have been. But there was very much a, 'oh well, A101 and A16 will make the tea and coffee', and 'A101 and A16 will put the chairs out and sort out the room', and everything like that.*

It is unclear whether this sentiment was a result of a gender bias, personality dynamics or because the voluntary capacity of the role of town councillor is unpaid and reduces the incentive to enact support functions, leaving those being paid to perform housekeeping duties. Thus, suggesting that without equal representation of women, in-group bias can affect intra-group solidarity and reciprocity. One respondent indicated that there was a natural process of being in a minority whenever a dominant group is prevalent within a group dynamic, suggesting that antagonism and conflict is an unconscious phenomenon.

This led to resulting perceptions and tensions along gendered lines and an unintentional gender bias against women as a minority group. Even when men (within IfF) were not intentionally patriarchal there was a sentiment that the structured presence of a majority over a minority gives rise to (unconscious) prejudices against the minority party:

> *A degree of [short pause] sexism that just springs from having mainly men … it's just that unintentional, you know, that kind of, the thing of if you're just in a minority, to start to get side-lined a bit.*[9]

In contrast, as more women were elected to become councillors in IfF's second administration, an IfF councillor marked how this change in representation transformed the "feeling" of interaction among them, as they suggested: "now we're, as a group, we're 50–50 … it just, it just feels different, the dynamic is really different".

To address this, an entity called 'Engaging Women' was set up by a female IfF councillor as a support network to encourage other women to "step up to the plate". Or as the group's Twitter group (in 2016) suggested, it exists for "Women who want to increase female participation and contribute to a more inclusive society. Be ready for a call to arms". Encouraging women's broader engagement in placemaking and public life was influenced by promoting solidarity amongst women and unlocking women's perceived inherent capacity to support one another. It is notable that Engaging Women has borrowed a similar governance model to IfF in the 2015 local election, where some women were encouraged to play either active or support roles (like how IfF's earlier election campaign was organised).[10]

Both IfF and Engaging Women share similar narratives of unlocking new energies into Frome's placemaking by increasing diversity—indicating the inter-organisational diffusion of governance characteristics through local political actor networks. Examples of women in political life at the national level were actively channelled through emerging local feminist networks in Frome to encourage other women to undergo role transformations and animate more (latent) feminine qualities of support and sharing, particularly for mothers who wanted to act as good role models for their daughters. By drawing in on the discourses of the increasing representation of women in politics in devolved territories (e.g., Scottish and Welsh party-political leaders), using feminist identities helped to weave in the progressive strands from other places to bring local favour and reinforce the identity of independence.

[9] They described how seeing the inequality between male and females led them to be "pi**ed off" and to unleashed their desire and agency to set up a local group to address this.

[10] Engaging Women continues to be a force for encouraging women's participation in Frome's politics—see https://www.frometowncouncil.gov.uk/quick-dirty-engaging-women-in-local-politics/

Conclusion

This chapter has explored how informal ties within Frome's placemaking networks influences the politics of engagement in transition arenas, or disconnection with them. We saw how the affective ties discussed in Chap. 6 cultivated interdependent and reciprocal multiple memberships (arenas) for relocalised entities in the town as animators of transition.

While the continual activity of the usual suspects sustained momentum and connection to placemaking issues, as such, they were felt to have a disproportionate influence on Frome's placemaking trajectories. The role of friendship and pre-existing ties was a key mechanism that created short-cuts into local government under IfF and was sometimes used as an active strategy to consolidate shared agendas for relocalised transitions. Indeed, a nucleus of these ties was perceived as either unconscious or active mechanisms of exclusion in the NDP process, where governance arrangements or the (over) representation of the usual suspects lacked legitimacy and a bottom-up (i.e., citizen) influence over the content of the plan.

There were several niche-regime broken links, notably on the idea of cohousing indicating that whilst these ties at the local level were strong and influential, this did not permeate into the LPA 'regime'. There were also niche-FTC links where role holders did not stabilise links between Sustainable Frome and FTC in the NDP process, causing unidirectional engagement with the plan. The plan also failed to sufficiently engage more 'ordinary' citizens who likely do not self-identify as being interested in planning and political action; this affects the legitimacy of relocalised action through community-based planning, suggesting a more inclusive participatory approach is still yet required to activate interest in planning among citizens more broadly. Recent developments in placemaking networks in the town offer potential ways to remedy this, such as the street-by-street approach of the FNN and its possible route to more inclusive, and potentially transformative, governance that can bridge placemaking and planning arenas in the next iteration of Frome's Neighbourhood Plan. (Though, as we have seen, this is greatly supported by resourced functions in the town council, such as the Community Development Projects Officer). In addition, as we will go onto explore in the remainder of this book, the urgency of the debates around climate change and the increasingly visible connection of planning instruments to respond to the climate and ecological emergency can also help to bridge some of the issues relating to participation and potentially develop a more inclusive agenda for resilient places and innovations in planning approaches. The challenge going forward is weaving together the innovative potential for planning to deliver on this while navigating broader 'placemaking transitions' in the context of political, demographic and environmental change, an issue we explore further in the next chapter.

References

Arnocky, S., Milfont, T. L., & Nicol, J. R. (2013). Time perspective and sustainable behavior. *Environment and Behavior*. https://doi.org/10.1177/0013916512474987

Belotti, F., Donato, S., Bussoletti, A., & Comunello, F. (2022). Youth activism for climate on and beyond social media: Insights from FridaysForFuture-Rome. *The International Journal of Press/Politics, 27*(3), 718–737. https://doi.org/10.1177/19401612211072776

Bourne, L. (2012). Report and feedback from the Citizen's Panels held in Frome from May–July 2012. Action for Market Towns. https://www.frometowncouncil.gov.uk/wp-content/uploads/2014/05/frome_report_from_citizens_panels_-_final_document_july_2012.pdf

Comer-Schwartz, K., Craig, B., Dixon, T., Gould, G., Jeffery, D., Mason, P., & Williams, K. (2023). A labour vision for community power: Participation, prevention and devolution. https://www.newlocal.org.uk/wp-content/uploads/2023/09/A-Labour-Vision-for-Community-Power-compressed.pdf

Franco, C., & Ghisetti, C. (2022). What shapes the "value-action" gap? The role of time perception reconsidered. *Economia Politica, 39*, 1023–1053.

Frome Times. (2023). Frome Town Council declares a housing crisis. *Frome Times*. https://www.frometimes.co.uk/2023/08/15/frome-town-council-declares-a-housing-crisis/

FTC. (2014). A neighbourhood plan for Frome 2008–2028 (v2). Frome Town Council. https://www.frometowncouncil.gov.uk/council/frome-town-council/planning-for-frome/fromes-neighbourhood-plan/

FTC. (2016). *A neighbourhood plan for Frome 2008–2028*. Frome Town Council. https://www.frometowncouncil.gov.uk/council/frome-town-council/planning-for-frome/fromes-neighbourhood-plan/

Håkansson, I. (2018). The socio-spatial politics of urban sustainability transitions: Grassroots initiatives in gentrifying Peckham. *Environmental Innovation and Societal Transitions, 29*, 34–46. https://doi.org/10.1016/j.eist.2017.10.003

Hickey, S., & Mohan, G. (2005). *Participation: From tyranny to transformation? Exploring new approaches to participation in development*. Zed Books.

Hutchcraft, J. (2021). "They kicked party politics out of the town": Spending the day with the punk rock mayor of Frome. *Kerrang*. https://www.kerrang.com/they-kicked-party-politics-out-of-town-spending-the-day-with-the-punk-rock-mayor-of-frome

Kyle, G. T., Jun, J., & Absher, J. D. (2014). Repositioning identity in conceptualizations of human–place bonding. *Environment and Behaviour, 46*(8), 1018–1043. https://doi.org/10.1177/0013916513488783

Lorenz-Spreen, P., Oswald, L., Lewandowsky, S., & Hertwig, R. (2023). A systematic review of worldwide causal and correlational evidence on digital media and democracy. *Nature Human Behaviour, 7*(1), 74–101. https://doi.org/10.1038/s41562-022-01460-1

Rydin, Y. (2012). Using actor–Network Theory to understand planning practice: Exploring relationships between actants in regulating low-carbon commercial development. *Planning Theory, 12*(1), 23–45. https://doi.org/10.1177/1473095212455494

Sabherwal, A., Ballew, M. T., van der Linden, S., Gustafson, A., Goldberg, H. G., Maibach, E. W., Kotcher, J. E., Swim, J. K., Rosenthal, S. A., & Leiserowitz, A. (2021). The Greta Thunberg Effect: Familiarity with Greta Thunberg predicts intentions to engage in climate activism in the United States. *Journal of Applied Social Psychology, 51*(4), 321–333. https://doi.org/10.1111/jasp.12737

Shelter. (n.d.). The cost of living crisis. https://england.shelter.org.uk/support_us/campaigns/cost_of_living_crisis

Wallis, H., & Loy, L. S. (2021). What drives pro-environmental activism of young people? A survey study on the Fridays For Future movement. *Journal of Environmental Psychology, 74*, 101581. https://doi.org/10.1016/j.jenvp.2021.101581

CHAPTER 10

Placemaking Transitions in Frome: Consequences of Independence and Relocalisation on Transformative Change

Transition and Relocalisation

Transition *and the Stages of Organisational Development*

The role of *Transition* on relocalised placemaking has been the core to this research, evaluating the change dynamics to steer particular visions of 'sustainability' in the market town of Frome, Somerset (speaking to the *Transition and Relocalisation* 'tribe' of the New Economy, depicted in Table 5.2). Here, we assess the effect of the occupation of the town council by the independent group, the Independents for Frome (IfF), on the local Transition Initiative in the town, Sustainable Frome. In this chapter, we further unpack how the dynamic of old and new has been recurrent—not only with the identity marker 'Old Frome' and 'New Frome' but within the organisational evolution of the two 'tribes' of the New Economy under investigation (relocalisation/transition and politics/democracy) which has affected relational ties and role transitions between IfF and Sustainable Frome over time.

As you may recall, we explored the context of Sustainable Frome's inception in Chap. 5 and presented a visual depiction of how the ties to *Transition* related to wider change dynamics in Frome and beyond. We take a deeper dive into the governance of these local (transition) arenas. Here, we critically explore the insight from a Sustainable Frome member I interviewed

A. Burnett, *The Politics of Transition*, Palgrave Studies in Sub-National Governance,
https://doi.org/10.1007/978-3-031-40769-7_10

who suggested the group had gone through key "stages" since its inception (see Table 10.1). In some cases, several key local transition actors weaved in and out as active members of Sustainable Frome (and national level environmental and political networks), with varying degrees to which Sustainable Frome was being "held" together. We see how Sustainable Frome itself was characterised as a parent organisation, helping to spawn and incubate nascent ideas, which then moved on and transitioned into entities in their own right, stemming from the Sustainable Frome "umbrella". At these various distinct stages (of transition), Sustainable Frome was seen to spin off in "a different direction"—particularly as some members left to become "more politically engaged" when Independent for Frome (IfF) councillors were elected in 2011 and former members of the Transition Initiative started to concentrate their activities for green and relocalised objectives from within Frome Town Council (FTC) and beyond, rather than Sustainable Frome as the primary conduit to do so.

In Stage 2, when IfF came into being and some active members of Sustainable Frome left, the Transition Initiative experienced an existential reflection. These key 'nodes' that now bridged Sustainable Frome and the town council (see Table 5.1) drew on shared (green) organisational identities and relationships. IfF now sought to utilise new placemaking and planning tools, such as the Neighbourhood Planning process and the rights underpinned by the Localism Act (see Chap. 7). This affected degrees of intra-group solidarity, reciprocity and cohesion (see also the body of literature around social identity theory, which has explored these issues in detail). As inter-group ties shifted, some actors foraying into new roles sought to differentiate themselves as protagonist political actors or spin-off autonomous entities in their own right while also seeking to retain ties to Sustainable Frome to various degrees (such as through multiple membership or drawing on initiatives some actors incubated in Sustainable Frome before becoming a town councillor, inlcuding the 'greener buildings' sub-group).

The rise of IfF was seen to reduce active membership of Sustainable Frome and shift the loci of environmental action in the town, where activities usually within the remit of social movements or local environmental groups were co-opted by a green-minded town council. Thus, the local environmental movement itself at this time can be characterised as being in a particular process of transition, so that what started as a more contained (niche) environmental remit moved increasingly into planning and more formalised roles. As a result, there was a sense that Sustainable Frome

Table 10.1 The stages of transition in Sustainable Frome

Stage/Date	*Summary*
Stage 1 (2006–2009)	Peter Macfadyen establishes Sustainable Frome independently of the Transition Network (since Sustainable Frome's inception pre-dated Transition Network) with a core group to shape meeting agendas and organisational direction. Both Sustainable Frome and the Transition Totnes (which the movement is well known for) were established concurrently - but independently - in 2006 (see Fig 5.1).
Stage 2 (2009–2011)	Key figures in Sustainable Frome put themselves forward into more politically active roles and were elected as IfF councillors; energy on green development channelled to FTC, the NDP and Frome Cohousing (a spin-off group).
Stage 3 (2011–2012)	No one "holding it" and apparent split between spiritual and practical-oriented people. The lack of formalised leadership led to tensions arising from those who wanted to have a "presence" in Sustainable Frome and its overall direction was weakened. Increased spiritual meetings, e.g., focusing on 'constellations' and mindfulness rather than appealing to more 'mainstream' elements of the Frome community (such as local businesses).
Stage 4 (2013–2016)	More established members of Sustainable Frome seek to re-orientate its organisational purpose to be more aligned within the broader environmental movement and (especially) *Transition*. Experimentation with governance technologies being pursued in the *Transition* movement, such as holacracy to support more autonomous decision-making. Some longstanding members drop out and a new tension arises with some emerging radical elements, e.g., campaigning against fracking. Many offshoots, such as Frome Energy Coop (FRECo) and Frome Cohousing, no longer formally associate with Sustainable Frome. Emergence of non-Sustainable Frome environmental initiatives (as a challenge and complementarity to Sustainable Frome as a key 'holder' of relocalised activity and identity).
Stage 5 (2017–2021)	The rise of Extinction Rebellion (XR) corralled the more radical/ campaign elements raised in Stage 4 and reflected the wider movement shift towards interest in direct action in the lead up to the Paris Agreement and the awareness raising of Greta Thunberg on climate issues. Some members of Sustainable Frome either left to become XR members or retained ties to both, but the distinct role between both groups needed to be worked out.

(*continued*)

Table 10.1 (continued)

Stage/Date	*Summary*
Stage 6 (2022–2023)	Peter Macfadyen increasingly worked with XR organisers to support a softening of its overall campaign strategy, moving from direct action to developing parallel democratic infrastructure at a local level for a more conducive and stronger grassroots democratic approach. For instance, strengthening the links between local climate governance, such as climate action plans and citizen assemblies in the wake of climate and/or ecological emergency declarations. Peter's wife Annabel (who played an active role in Sustainable Frome and Frome XR) also becomes a steering group member of the Frome Neighbourhood Network (FNN) to support resilient street-by-street networks. FNN champions the Transition Streets model (advocated by Transition Totnes) as a vehicle to strengthen localised climate and well-being responses in communities (see Chap. 9). Other relocalised, sustainability frameworks—such as Kate Raworth's Doughnut Economics (Raworth, 2018)—are starting to shape Frome's environmental and political discourses, with a 'Doughnut for Frome' being developed, in collaboration with Sustainable Frome.

had nothing to push against (see Böhm, Dinnerstein, and Spicer, 2010). In part due to the social capital and ties of action when the "movers and the shakers" started to work within state structures, an IfF councillors relayed how local groups initially "took their foot off the pedals" and there was a "big vacuum" that was left within Sustainable Frome, which, according to one member, affected energy flows and momentum to "keep up steam". This had several internal consequences, including a sense of questioning Sustainable Frome's overall purpose as it struggled with its own role transition. IfF was seen to have "muddied the waters" of responsibility on delivering on the community priorities for the Community Plan and the green agenda (see Chap. 5):

> *What is the town council's responsibility, what these different pressure groups are actually pressuring for and against which is harder to define. And it actually makes, erm [pause] it makes it harder to get a head of steam up [laughs], sometimes [smiles]. Erm, sort of a victim of our own success. Sometimes it's easier to push against a barrier than just to move forwards because it's been taken away.* (Sustainable Frome member)

> *I assumed that Sustainable Frome and the other groups ... environmental groups would come to the council and say this is what we want to do, can you give us money? And I think what they thought—seeing ... all [the] sort of main people within Sustainable Frome [had transitioned into FTC] was that we'd be doing everything. So actually, they took their foot off the pedals and Sustainable Frome did too, and as I dropped out completely from Sustainable Frome it went into hiatus a bit to be honest ... It's got potential but it's a bit, they're kind of working it out. It's definitely a bit flaccid*[1]*... The monthly meetings happen. So, all that's happening. And there would have been 20–30 people there last, last night. So, you know in one sense it continues.* (IfF councillor)

These changes hollowed out a sense of group purpose and led one Sustainable Frome member to seriously question the role of the organisation at one particular juncture, with an existential reflection of "looking at our navel quite a lot saying what is our role now?". This reflects Melucci's observation that social movements need to choose how they act but that in some cases, external forces can—temporally—take this decision-making power away from them until they reassemble their sense of mission in a changed context (see Melucci, 1989):

> *If you're not a pressure group, sort of like writing to the local council and to Mendip council—if you're not a focus group, if you're not a consultee, what are you, other than some kind of social network? (Sustainable Frome member)*

Without such a push-pull driver, relative satisfaction with (local) state actors (the new IfF councillors) affected motivation and the excitement to be committed to a cause in a given organisational container (i.e., a transition arena made up of actions by Sustainable Frome). Indeed, the challenge of role transition was apparent not just for Sustainable Frome but for some other placemaking entities in the town who withdrew a more oppositional or campaigning stance against the town council due to a relative "satisfaction" with FTC under IfF (before IfF, FTC was perceived to not take sustainability seriously enough). Moreover, ties between remaining Sustainable Frome members and the town council were often more emotive—"feeling at one" with FTC in their shared inter-organisational objectives—rather than pro-actively attending town council meetings (again echoing the role of personal and affective ties in forging inter-organisation

[1] This term was used response to my question on whether the organisation had become more "flaccid" and is not the respondent's own phrasing.

networks). Without an external push factor or resistance (e.g. contestation and oppositional identities), there was a partial breakdown in the activating conditions of engagement between Sustainable Frome and FTC; this morphed into engagement between these entities being shaped perhaps more by self-interest (as raised in Ch 9) on a more unilateral basis, e.g. through individual members of Sustainable Frome who enjoyed in getting engaged in planning consultations or reading planning documents.

While *Transition* has strongly prefigurative (i.e. the real-world demonstration of intentional practices) and relocalised objectives, in Stage 3 its degree of philosophical openness and respect for difference without a political or campaigning approach was considered a risk which opened entry points that led to competition over its purpose. Because of the broad spectrum to which Sustainable Frome appealed (i.e., local food, local energy and renewables, personal 'inner' transitions and resilience), there was a trade-off between influence and values in an organisation that appeals simultaneously to esoteric and practical interests. During this stage, the "more spiritual people" sought a greater profile for issues important to them. One respondent cited email correspondence that suggested an 'outsider' from a particular religious denomination wanted to have a "presence" in the group (apparently acting opportunistically as members appeared to no longer hold the Transition Initiative together). There was a subsequent split between spiritual and campaigning elements within the organisation as well as tensions over appealing to the local business community (e.g., economic relocalisation) and other facets of Sustainable Frome's more spiritual identity and practices (e.g., personal 'inner transitions').

Many of those who came together in (pre-)Stage 1 of Sustainable Frome revolved around a friendship network with countercultural and environmental values. This included the Macfadyens and several others who later took on more formalised and political roles to advance sustainability issues in the town. This particular cohort spoke about how they attended "hippy camps" locally and became close friends through these emerging *Transition* arenas.Hoewever, many of these people had mainly disengaged from Sustainable Frome as the organisation moved into Stage 3, including some that actively withdrew to become more political (via IfF) or who stepped away to enable new growth in the organisation. There was a sense of previous networks fragmenting, with old friends too busy to see each other or individuals evaluating their own commitment to the organisation; some members felt they had given enough of themselves to the group in holding it together and risked personal burnout, owing to not being appreciated or sufficiently thanked for volunteered work, undermining reciprocal ties in the organisation.

In Stage 4, a challenge for Sustainable Frome was to fill various roles after others stepped down from 'leadership' roles. Under the guidance of one Director, the organisation used holocracy—a governance technology being experimented with in the wider *Transition* movement—to help "break [Sustainable Frome] open" owing to a suggestion from Transition Network and known experimentation with the process by the neighbourhood Transition Initiative in Glastonbury. As a Sustainable Frome member explained, holacracy is a process of "formalised anarchy ... a *structured* process for making decisions in an organisation" which gives guidelines to actors to help them perform their roles with high degrees of autonomy. Using holacracy aims to integrate an empowering philosophy to call out organisational tensions to help get people "unstuck", such as by bringing concerns to meetings to resolve particular issues with an agenda that is set by those present (see Kumar and Mukherjee, 2018).

However, the experience of using holacracy was perceived to have several steps that needed to be learnt, which some respondents suggested were too technical and took the fun out of meetings. Meanwhile, setting an agenda at the meeting—rather than in advance—could be exclusionary for those who couldn't attend to shape the agenda at the time. Holacracy was seen as unable to fulfil roles, only identify a *need* for them. Thus, indicating a tension between steering more 'emergent' qualities of organisational change. Moreover, the loose nature of Sustainable Frome called into question the qualifications for participating in holacratic meetings (i.e., should it be for all members on the mailing list, directors, monthly meeting attendees, etc.).

Sustainable Frome underwent yet another shift in Stage 5 as interest shifted to direct action advanced by the Extinction Rebellion (XR) movement, particularly around the time of the Paris Agreement and a growing recognition of the need for radical climate action to be taken. This created, yet again, pressures in the governance of Sustainable Frome as some members sought to channel their energies in the more campaigning orientations and identity of XR, with Sustainable Frome once again acting as a holding ground for broader environmental action in the town. (Some IfF town councillors were among the Frome XR activists arrested in XR demonstrations—again symbolising the radical approach to Frome's (environmental) politics and the continued links between IfF and environmental movement activism (see Table 5.1)).

When I caught up with Peter Macfadyen in June 2023, he suggested that there had been yet another shift in XR-Sustainable Frome relationships. During Stage 6, XR sought to change tack and downplay its direct action towards more conciliatory approaches, working in partnership with local organisations, such as Sustainable Frome. Though, what is even more interesting is XR's revised strategy to circumvent political institutions and cultivate a demand for climate governance through the ordinary engagement of all local groups beyond those typically seen as playing an active role in environmental or political life. For instance, encouraging greater interest in environmental action, craftivism (using arts and creativity for symbolic protest—such as by enrolling the local knitting group) and seeking to awaken civic interest in the environment through participatory arenas, namely local citizen assemblies. Peter Macfadyen's wife Annabel—also an active member of Sustainable Frome and XR with a keen interest in well-being in the town—is also a member of the Frome Neighbourhood Network (FNN): a street-by-street approach to connecting and empowering communities to self-determine issues important to them and their solutions through a place-based and digital peer-to-peer network (see Chap. 9). This marks another shift in the stages of Frome's relocalised governance in ways that link the town to broader New Economy trends (such as the sharing economy) in the context of digital democracy and a bridging of the tribes of *Transition and Relocalisation* (via Transition Streets and the FNN) and *Government and Politics* (via XR).

Dynamics of the Parent and Child in Sustainable Frome's Stages of Transition

As various spin-off groups emerged from Sustainable Frome over time this resulted in interesting degrees of alignment (and conflict) with 'roots' to transition (see also Chap. 5). According to a Sustainable Frome member, these emergent entities stemming from the Sustainable Frome parent node were meant to use its name in the marketing material of these new spin-off groups to recognise their roots in *Transition* in their marketing material. This was an agreed principle established by, but one Director suggested that this was rarely done in practice. (This has interesting parallels with the emergence of the 'new' and 'old' in the recognition of the role that Frome's Community Plan played in shaping Frome's Neighbourhood Plan; as depicted in Fig. 9.2 various groups were mentioned that—to varying degrees—contributed to the relocalised flair of Frome's plan but the Community Plan (Vision4Frome) Management Group was overlooked.)

Moreover, the lines of communication between spin-off groups and Sustainable Frome have not always been reciprocal, with one respondent citing how one spin-off group didn't inform them that they had also voted to use holacracy in their meetings. That these spin-off groups replicated what Sustainable Frome did was characterised as organic and akin to a parent-child relationship, where the young seek (and at times struggle) to break free from the family nest and find their own autonomy, indicating strongly emergent and path-dependent qualities between relocalised groups centred around Sustainable Frome and *Transition*.

However, attention to governance and holacracy meant one member of Sustainable Frome felt they had taken their eye of the purpose of the organisation. This then gave the space to encourage other action-orientated groups to establish—almost as counter groups (like XR) for "people who find Sustainable Frome a bit boring" or as a "talking shop" (words used to describe Sustainable Frome). There was a sense that different clusters of initiatives incubated within Sustainable Frome were becoming increasingly fractured. As one member suggested, this particular stage led to more notable differences between members which became "more apparent than the things we have in common".

Nevertheless, the role of close interpersonal affective ties and the protective boundaries of the organisation as an arena for learning and sharing, or as one member described, "the importance of social gluing people together", helped to sustain Sustainable Frome through its own transitionary period; again indicating how informal bonds act as a conduit to link to broader networks and sustain organisations (which has been described in various papers written on social movements, e.g. Yu, 2012, see also Funke, 2014). As we saw in Chap. 5, close personal ties that were important to Sustainable Frome's inception had also travelled into the spaces of government. Some IfF councillors were either former or current members (Table 5.1), with one acting as an informal liaison between Sustainable Frome, and the FTC the Resilience Officer (later Manager—a role that began initially as 'Energy and Recycling Officer'), was also a member and director—ties that, as a Sustainable Frome member suggested, meant FTC had "been really helpful in supporting us". (As we will see in Chap. 11, former Sustainable Frome members also later came to be elected representatives at district and unitary council levels with wider implications for the scaling up of green politics).

Though, not all of spin-off groups had success at the time of the interviews. The research revealed the presence of latent initiatives, where organisations wound down their activities to play a dormant role, should they need to be resuscitated. For instance, owing to the lack of progress in acquiring a site to build a cohousing community because discussions on this were "stuck in planning" or hampered by conflicting priorities of landowners this led Frome Cohousing to "wind down" their activities (and later reasserted itself in the *Mayday Saxonvale* campaign). Keep Frome Local (Saxonvale) and Vision4Frome (Community Plan) members also described a similar 'holding operation' stance. (And, indeed, the *Mayday Saxonvale* campaign arguably became a more recent iteration of the Keep Frome Local campaign—see Chap. 8). (The issue of latency in community development is a long-standing concept, see Young and Mayo, 1959. For a discussion on the role of latency in social movements, see Saunders, 2007).

Nevertheless, the strongly interlinked environmental organisations in the town (see Chap. 6) were a mechanism to ensure the longevity of pushing for development alternatives and planning for transitions, even if a future FTC administration made posts like the Resilience Manager redundant. Another Sustainable Frome spin-off group member spoke about their underlying anarchist spirit which could be activated if the far right emerged as a political force where "I'll be out on the streets with my pitchfork, anarchist's headband [on, laughs] [if the far right got into power]"—which seemed a slight possibility during heated rhetoric on Brexit. These underlying, latent propensities to engage in more activist or relocalised future behaviours were incubated in particular contexts but became powerful tools that could be enacted as future conditions changed. This suggests that the transition of role boundaries sometimes 'stops' to allow other phenomena to settle but these temporal elements are part of the 'holding together' of the momentum for change that lies dormant in communities until these sentiments once again become activated.

Bends, Splinters and Fractures in the Green Movement

A respondent who left Sustainable Frome to become an IfF councillor in 2011 suggested their overall impact was much greater in FTC than acting within *Transition* alone, indicating these actors could achieve more *within* government than outside—either by setting up initiatives in the town council, being strongly linked to or facilitated by it, or securing financial assets or other resources for local groups. (Indeed, IfF's independent political approach was endorsed by one *Transition* actor as "the only way" for the movement to become more politically engaged). Thus, indicating how relocalisation reached a degree of ***maturity*** and power within a different organisational setting (the state). Indeed, one respondent questioned whether IfF's occupation of power and success was indicative of *Transition's* purposeful demise when relocalised ideas become mainstream (i.e., steering for emergence)—or indeed the maturity of those involved in it. Another view was that this journey was just a "bend" in Sustainable Frome's history and the shared goals of relocalisation in the town, indicating the intersecting qualities of steer and emergence in a complex network of relocalised initiatives.

However, despite Sustainable Frome's deep networks in the town, there was a sense that *Transition* alone wasn't sufficient to bring the transformative change it sought. Some initiatives supported by Sustainable Frome were recognised by one member as not being able "to change the world … [but] it sort of raises the possibility". However, the role of Transition Initiatives was questioned by an FTC representative who suggested that *Transition* was like the "canary in the miner's cage" on environmental issues or that their "impact" in the town was not visible to them. Some respondents questioned the nature and scale of the change *Transition* promotes, especially if it takes a less political role where the movement lacked the tools to steer effectively (see Chatterton and Cutler, 2008).

Another suggestion was that environmental action would only be scaled up with radical policy solutions, particularly to break the cycle of prioritising economic over environmental issues or only caring about the environment during cycles of economic productivity. However, these reflections point back to a comment raised by an FTC staff member that alternative development models (like cohousing) purporting to the values of *Transition* were like "fiddling while Rome burns", while on the other hand alternative development models within the New Economy can help to change the "frequency" of development ambitions.

At a movement level, there were signs of fracture as *Transition* found itself within a wider "spectrum" or ecosystem of potential strategies to achieve change, competing at times with different organisational identities in the environmental movement scene, such as those that sought direct action being amplified through the rise of XR. (Some people I interviewed felt that *Transition* had repackaged existing ideas from other environmental social movements and made claims to them as their own—such as visioning methodologies or open space—suggesting, like the examples discussed earlier, that in an attempt to make its mark it had not paid sufficient homage to its path dependent roots.)[2] A Sustainable Frome member suggested that at a movement level there was a wider point of transition, where people were starting to fall into "camps" to differentiate their positionality within the (New Economy) movement ecosystem and that *Transition's* stance of "trying to be all things to all people" has attracted varied perspectives but at the same time has led others to "drift away" (exemplified in the way Sustainable Frome's organisational purpose was subject to contestation between different factions of its membership base).

Stage 5, the rise of XR, was once again seen to be "taking a lot of energy out of Sustainable Frome", as IfF had done. Peter Macfadyen described himself as "boxing between the two" environmental groups with differentiating identities broadly characterised as "hippy" (Sustainable Frome) and "way too radical" for some (XR).

The overall transference between different factions of the environmental movement (such as *Transition*, XR and green political vehicles) was tied to degrees of crossover of personal values, which configured multiple membership to different groups. Jumping on the "*Transition* bandwagon" (i.e., ascribing to alternative development principles, permaculture or 'inner transition') was not deemed to be a seamless process. There was a feeling that lines are drawn between personal values and organisational or social movement ones resulting in tensions as individuals navigated the most appropriate organisational loyalties according to their own value compass, allowing people to both "jump on" and "jump off" environmental organisation identities (see Hatch and Schulz, 2002). As a Sustainable Frome member suggested:

> *I'll happily jump on the Transition bandwagon, if it's something that works. As long as it's not compromising my principles too much, more than 5%. I'll compromise a bit, you know, but not too much.*

Indeed, in August 2018 the town's local newspaper the *Frome Times* suggested that the future of Sustainable Frome "hung in the balance after a shortage of volunteers" but now had a "bright future". Those attending this meeting expressed their support for the group and its value in raising sustainability issues in the town. Akin to Rob Hopkins'—the co-founder of the *Transition* movement—book (Hopkins, 2019), the phrase '*What*

[2] Such as the practice of visioning and utilising ideas of sustainable development stemming from the 1970s.

If was used as a rhetorical question to imagine how to invigorate sustainability issues in the town and has been used as a powerful activating and reflective discourse within the broader *Transition* movement, Sustainable Frome, FTC-facilitated or Sustainable Frome spin-off groups, local groups such as Fair Frome Housing, and in IfF's own messaging and campaigning—i.e., 'What IfF'. The Frome Neighbourhood Network (FNN) has also asked the question "What if you and your neighbours could change the world?" to stimulate ideas based on its 'Neighbourocracy model' which aims to organise local democratic mutual aid (see Frome Times, 2023). This reflects how creative capital can be unleashed through specific phrases that stimulate collective imagination, in what sustainability transitions scholar Flor Avelino has termed *innovative* powers, that is, 'dreaming' discourse and practices that have the power to re-imagine and agitate change (Avelino, 2017).

At the time of my original interviews (Stage 4), action on the environment was felt to have moved from eco-fatigue and disempowerment towards a need to be more connected and recognise shared interests across the broader environmental movement (see Stage 6, Table 10.1); doing so was felt to unearth a renewed global social capital for collective responsibility on environmental action, based on trust and reciprocity to challenge self-interest and short-termism in the use of resources (i.e., steering for collectivism).

More recently, there has been increasing interest in regenerative development (see Burnett, 2022a) and a move towards the sharing and the circular economy. This brings together various ideas within the wider New Economy movement, ideas which are increasingly mainstreamed across different business sectors and community, and government. For instance, in Frome in November 2023 a new informal discussion group was set up that aims to meet regularly to develop a 'Doughnut for Frome', based on Kate Raworth's concept of Doughnut Economics, which seeks to focus action around recognition that we live within finite planetary boundaries. The doughnut model is gaining global traction as a framework to help shape discussions on sustainability and is also being used to shape planning and local government decision-making (see Burnett, 2022b). However, as James (2022) highlights, caution is required so that the popularity of circular economy and regenerative discourses do not perpetuate neoliberally framed relationships between economics and ecology domains. For instance, Turner and Wills (2022) suggest that in efforts to 'downsize the doughnut' it is essential to always pay critical attention to the fact that such attempts exist within these complex systems; there is a danger these may become co-opted into yet another technocratic pursuit that dampens the regenerative potential of sustainability models to reflect local needs and unique place-based identities, such as those raised in this book.

Standardised Identity Tokens

As we saw in Chapter 5 and outlined in Appendix A, there are many instances where Frome's various interconnected environmental, social and political and economic enterprises share similar visual identity tokens that reinforce a sense of a coherent identity in the town. The visual twinning between the broader activist environmental movement and local political and placemaking approaches is reflected in Fig. 10.1. Here, we can see how XR Frome's visual organisational identity is aligned to the

Fig. 10.1 Formalised visual identity twinning of XR and FTC logo font (XR Frome, 2019)

standardised 'Frome' font used by FTC in its logo (see Fig. 4.2), rather than Sustainable Frome's (despite many members of Sustainable Frome being involved in XR). This is also perhaps indicative that XR is a UK-founded international environmental movement and of a different scale to some of the more local, relocalised entities that have spawned from Sustainable Frome and its networks. Other notable examples of the visual and discursive alignment of identities within Frome's placemaking and political networks can be seen through the use of word 'independent', and 'flatpack' in EdSpace's tiny homes marketing. EdSpace was another initiative created by a former IfF councillor (and former member of Sustainable Frome and Frome's Neighbourhood Planning steering group), EdVenture and the entrepreneur (who set up Forward Space co-working hub, the Frome independent and purchased of many shops on Catherine Hill to support creative, alternative enterprises). It reveals how small networks of key individuals in the town actively shaped place-making identities and alternative development models through shared discourse and visual identities, suggesting the reinforcive qualities (see Avelino, 2017) of niche-directed transitions as they sought to stabilise themselves in the fabric of placemaking networks key to expressing Frome's relocalised and alternative placemaking networks visual and cultural coherence. Similarly, more recent examples of utilising particular discourses visually include the FNN use of the Community Plan's alias, 'A Vision for Frome' in depicting the group's 'Our Vision for Frome' (see Fig. 10.2) which sets outs its goal to achieve a "caring, sharing, well-connected society where everyone matters and all voices are heard". (A vision which indicates the fusion of New Economy principles and ultra-local democratic participation stemming from both Annabel and Peter Macfadyen's activist agenda, amongst the many others involved in these pursuits in Frome).

Demographic Transitions and the Role of Intergenerational Dynamics

Personal and demographic transitions and shifting role boundaries were fundamental to these broader organisational shifts in the relocalised networks in the town (pointing to the assertion that the individual is a key element in any study of broader transitions, e.g., Pesch, 2015; Avelino & Wittmayer, 2015; Rauschmayer et al., 2015).

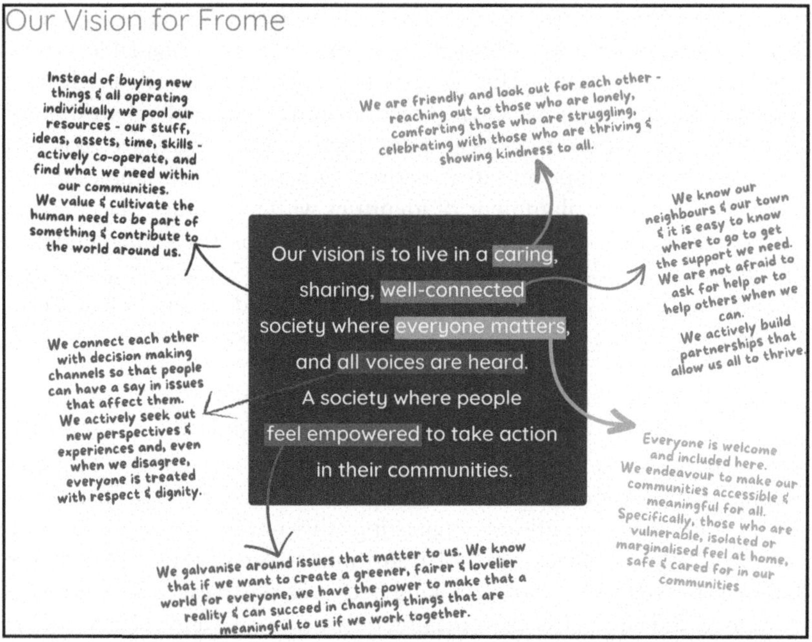

Fig. 10.2 A renewed Vision for Frome (Frome Neighbourhood Network)

Homophily—that is, the desire to be with people 'like me' (for an excellent discussion on this see McPherson et al., 2001)—was an emergent quality for the town in gaining an increasing reputation for being "green and trendy". A shifting demographic was perceived as a "rolling stone" of New Frome incomers who were attracted to Frome's green reputation, rather than necessarily bringing that experience with them to craft its cultural identity in the same way as the earlier cohort of Sustainable Frome might have done explicitly (because it was already 'there').

More recently, changing consumption habits and affordability issues were influencing the types of people attracted to the town. With a housing market that continues to push out those with modest savings and there are fewer opportunities for entrants on the housing ladder to buy property in Frome (i.e. there are less properties that can be "done up" or are be affordable for those on a lower salary), as with many other places across the UK. This affects the nucleus of people, or the dynamics, that corral certain identities to a town. Whereas some transition members who have lived in Frome for the past 20 years or so had a shared, often activist, environmental history, the consolidation of this identity has caused a more gentrified cohort effect, though still attracting an "outwash effect" from nearby Glastonbury, known for its alternative spiritual practices and world-renowned music festival as well as the attraction of living in Frome itself.

Finding a suitable 'place' to retire or bring up a family were massive drivers in wider demographic changes and personal role boundary shifts around new life-course priorities. Respondents differed in their approach to ageing in a more mature stage of life; retirement was felt to free up time for people to engage in placemaking as they could be more committed than others who were working while some didn't take on any new roles. This was due, in the words of a local group representative, to a desire to wind down to seek "quality of life" and to "create space" for personal pursuits and actively disengaged in multiple placemaking roles, sometimes owing to ill health (mainly from 'getting older'). (See also Burningham & Venn, 2017 for an interesting discussion on the role of life-course transitions on sustainable consumption patterns.)

Age and intergenerational interests were found to be of particular import in framing new development models to give hope and meaning to the next generation, indicating potential generational disparities in how environmental or development issues are responded to. Reflecting on how the maturity of Sustainable Frome as an organisation reflected the leadership qualities or personal circumstances of those involved in holding it together led one Sustainable Frome member to question whether the organisation was "just a husk, a parent growing old, in an old house". Recognising that the group needed new energy, they asked whether the baton of sustainable action in the town ought to pass onto the younger generation—as a form of retirement and empowerment (i.e., *Transition's* own intentional demise to graciously step away when its met its goals to influence a more sustainable future); thus reflecting an intergenerational transition of responsibility, where "I'm going to let you go off and do that now and I'm going to go off and look after myself".

However, to do this requires an active and engaged younger network to pass such a baton onto that can take advantage of such windows of opportunity. One younger radical protagonist who sought to "agitate" changes in the development process in Frome and beyond suggested that age disparities prevented them accessing and influencing Frome's relocalised identity network, even if they shared similar values. Though an IfF councillor suggested this might have been more to do with their capacity to communicate their value proposition to the town council; they also didn't live in Frome which, given the strong role that being from and understanding Frome was in Frome's Neighbourhood Planning governance may be another condition that prevented them from accessing the town's tight knit relocalised placemaking and planning networks.

Age disparities were also a reason for older people not being able to grasp the progressive agenda, within either the incumbent (pre-IfF party-political) FTC administration or more "Conservative" wards on the "dormitory estates" that were seen to be more divorced from energies of the town centre. This goes back to the oft-mentioned assumption that older people are harder to 'break' in terms of taking on new values or pursuits—implying that being older brings with it a more conservative mindset or being less able to take on new ideas (and with that a propensity for them to less likely be adopters of ideas like *Transition* or jumping on any particular environmental or political movement bandwagon 'if they're not that way inclined'.

There were clear influences of intergenerational factors influencing people's understanding of their engagement in transition arenas, shaping how they performed various roles or why they might be attracted to live in Frome in the first place. For instance, the influence of parents' professions steering one's own professional choices or having children as a reason why people were—or not—prepared to put their heads above the parapet. Being from the South West English region or having a partner or spouse from the area led many to make the move to be closer to family. The presence of the Steiner school—a child-led, alternative pedagogy developed by the late Rudolf Steiner (and referred to by one respondent as a bit of a "cult" they sought to distance themselves from)—was reported by one respondent to be a key conduit for new members to both move to the town and become members of Sustainable Frome (due to similar alternative value orientations). Indeed, at the One Planet Sunday event I attended in September 2014 at the Cheese and Grain, there was a Steiner teacher on the door to welcome people and many attendees were families attending the school. Thus indicating shared values and affective ties towards the New Economy agenda between the democratic and sustainability aims of the One Planet event and illustrates how informal civic duties become shared through representation in different relocalised groups. However, the school was taken over by a trust due to poor results by the school inspection agency OFSTED (with the entire academy of Steiner Schools in the area being awarded 'inadequate' and concerns over pupil discipline and how incidents of bullying were dealt with). Nevertheless, many former parents and children of the Steiner School became embedded in Frome's fabric of alternative placemaking and also became later members of XR.

However, despite the clear 'alternative' values that are readily expressed in the town, more research is required on the underlying divisions - or seeming divisions - between 'Old' and 'New' Frome on class-based lines and how different markers of identity might affect how the town evolves into the future. For instance, the continued high levels of anti-social behaviour in the town indicate this may be a fault line that may fracture the current trajectory of transitions in Frome, where there are frequent incidences of small-scale social unrest reported in the local media (refer to discussions in Chap. 8 for more discussion on the role of class in the politics of Frome's placemaking networks).

Physical Changes to the Town's Visual Identity

In Frome, there is evidence for the links between the town's increased popularity and changes to the physical and demographic landscapes in the town, sparked in part from the occupation of (economic) relocalised entities such as Forward Space, the Frome Independent and Catherine Hill. Physical urban landscape changes include a former discount store at the bottom of Catherine Hill becoming a restaurant, and the town council offices have been upgraded from their Palmer Street setting to the more majestic former Frome Rural District Council building on Christchurch Road; a move that has enabled the town council to expand its staff significantly as well as offer several community groups that use facilities and office space, including the local radio station, Frome FM. It also indicates how the historic fabric of a place can be refashioned and repurposed to give new life to the development of the future of place. For instance, the Silk Mill is a local (unlisted) heritage asset that was rescued from disrepair and possible demolition and its historic qualities are now intertwined with the creative and cultural capital rooted in the town. Similarly, the Cheese and Grain - another historic asset - was also upgraded and plays a key role as a venue that sustains the town's music, creative and social scene. Thus reflecting the permanence and strong conditioning function that Frome's historic environment plays as a durable facet of the town's place identity (see Næss, 2016) - which cannot be replicated and therefore utilised in the same way as other towns seek to replicate Frome's relocalised or Flatpack approach.

The Zeitgeist and Fertile Soil

One IfF councillor suggested the attempt to capture local political institutions and advance relocalised politics was considered to be in tandem with a window of opportunity. Driving forces had begun to change the economic nature of the town due to the actions of established local actors as well as, in the words of an IfF councillor, the move of more "young and zappy" New Frome voters "who were more likely" to vote IfF. Thus, suggesting that IfF, as one IfF councillor recounted, took advantage of a "moment of collapse" and that the "Zeitgeist was already there". Serendipity, luck or a more "universal" driver of fate and circumstances in Frome was also seen as a reason for why events took place and played a role in enrolling what one local group representative called "a reasonable thread of professional connoisseurs" to help advance organisational objectives for environmental groups. Thus, indicating how strategic relocalised initiatives at the local scale tapped into salient, emergent features of the New Economy. At the same time, *Transition* actors suggested that Frome had more "fertile soil" or it was "a really fertile space" for achieving the conditions for (relocalised) transitions:

> *You can throw soil onto fertile land or unfertile land sort of thing.* (Sustainable Frome member)

As a result, Frome's New Economy enterprises were felt by an IfF councillor to have reached a social "tipping point … [where] they can sort of work off each other" in relocalised economic transition arenas, or as a Sustainable Frome member suggested, this now makes various strands of activity less easy to pin down and determine where the roots of particular activities lie. However, as we discussed in Chap. 8 planning policy and land availability were still repeatedly seen as the principal constraint for the expansion of these types of activity taking root even further. Moreover, the administrative configuration of planning and politics circumvented the potential to replicate transition processes from other countries, notably Europe. For instance, one respondent suggested that an architectural style common to some European countries to build more narrow houses with basements could address land use pressures; however, the topography of Frome (as relatively undulating) was seen as a factor as well as the planning policy context for why this model couldn't be replicated.

Attracting Outside Attention

As one New Economy actor suggested, Frome's *Flatpack* approach is acting as a magnet for organisations "on the [same] radar" to collaborate, replicate and amplify Frome's experimentation with political and development alternatives. Frome is also attracting other relocalised initiatives to the town, such the 2019 annual networking event of the Dorset-based alternative magazine Stir to Action annual networking event that was marketed as a "playground for the new economy", acting as a site of experimentation and prefiguration towards shared values and practices (STIR to action newsletter, May 2019[3]). This is also depicted by the continuation of various national and local media articles on the popularity of the town in terms of its trendiness. For instance, in 2019 a Sunday Times headline suggested that Frome, yet again, was one of the coolest places to live in the UK (see Chap. 4) and noted "move over, Brixton—these are the coolest neighbourhoods in Britain. Get in before they lose their edge". This is an obvious example of the role of the media in stimulating patterns of gentrification by encouraging people to "get in" (i.e., buy property) in places like Frome before the impact of such actions leads to a dilution of their "edge". This suggests how place-based cultural creativity linked to social movements such as Sustainable Frome can easily become commodified (see also Håkansson, 2018).

Conclusion

The chapter highlights the dynamics and tension between old and new, including the parent-child relationship between Sustainable Frome and its associated spin-off groups, the perceived disparities in mindset by the younger and older generations, and the draw of family influences as important drivers of role transitions in development(al) transition arenas. For instance, momentum for change was crafted by relocalised entities to broaden the appeal of the green agenda, although this was at the expense of the energy within Sustainable Frome, which experienced an existential reflection when key local innovators moved into IfF and a resistance and contestation by green protagonists in local transition arenas was dampened. The breakdown of activating conditions (e.g. contestation and

[3] See: https://www.stirtoaction.com/magazine

oppositional identities) for some members of Sustainable Frome thus further hollowed out inter-organisation activities where 'sympathies' to a cause took over from showing up in town council spaces and moving into more formalised arenas, such as town council meetings. A splintering of local relocalised networks and fractures within the wider green movement were seen as wider forces influencing the trajectory of *Transition*, particularly if the movement was perceived to play a marginal role in influencing policy solutions or seeking too broad an appeal.

Meanwhile, the elevated appeal of the town to New Frome was linked to homophily but also a change in the way Frome's cultural identity was reproduced as different economic drivers influenced the socio-economic profile of those moving to the town, raising tensions about class and gentrification in transitions processes. In the next chapter, we examine how IfF has shaped and potentially transformed the politics of the town and invited in new forms of political—independent—identities and what this means for the politics of transition.

References

Avelino, F. (2017). Power in sustainability transitions: Analysing power and (dis) empowerment in transformative change towards sustainability. *Environmental Policy and Governance, 27*(6), 505–520. https://doi.org/10.1002/eet.1777

Avelino, F., & Wittmayer, J. M. (2015). Shifting power relations in sustainability transitions: A multi-actor Perspective. *Journal of Environmental Policy & Planning, 18*(5), 62–649. https://doi.org/10.1080/1523908X.2015.1112259

Böhm, S., Dinnerstein, A., & Spicer, A. (2010). (Im)possibilities of Autonomy: Social movements in and beyond capital, the state and development. *Social Movement Studies: Journal of Social, Cultural and Political Protest, 9*(1), 17–32.

Burnett, A. (2022a). Operationalising performance measurement and incentives for 'regenerative value' to improve net sustainability outputs. In G. Vélez-Cuartas & O. Romero-Goyeneche (Eds.), *Transformative metrics: Contributions to the studies for monitoring and evaluating how science, technology, and innovation can address social and environmental challenges.* University of Antioquia.

Burnett, A. (2022b). Securing a regenerative and just planning system. *Town & Country Planning, 91*(2), 123–128.

Burningham, K., & Venn, S. (2017). Are lifecourse transitions opportunities for moving to more sustainable consumption? *Journal of Consumer Culture, 20*(1), 102–121. https://doi.org/10.1177/1469540517729010

Chatterton, P., & Cutler, A. (2008). *The rocky road to a real transition: The transition towns movement and what it means for social change.* Retrieved from http://trapese.clearerchannel.org/resources/rocky-road-a5-web.pdf

Funke, P. M. (2014). Building rhizomatic social movements? Movement-Building relays during the current epoch of contention. *Studies in Social Justice, 8*(1). https://doi.org/10.26522/ssj.v8i1.1037

Hatch, M., & Schulz, M. (2002). The dynamics of organizational identity. *Human Relations, 55*(8), 989–1018.

Håkansson, I. (2018). The socio-spatial politics of urban sustainability transitions: Grassroots initiatives in gentrifying Peckham. *Environmental Innovation and Societal Transitions, 2018*(29), 34–46.

Hopkins, R. (2019). *From what is to what if: Unleashing the power of imagination to create the future we wan.* Chelsea Green.

James, P. (2022). Re-embedding the circular economy in circles of social life: Beyond the self-repairing (and still-rapacious) economy. *Local Environment, 27*(10–11), 1208–1224. https://doi.org/10.1080/13549839.2022.2040469

Kumar S, V., & Mukherjee, S. (2018). Holacracy–the future of organizing? The case of Zappos. *Human Resource Management International Digest, 26*(7), 12–15.

McPherson, M., Smith-Lovin, L., & Cook, J. M. (2001). Birds of a feather: Homophily in social networks. *Annual Review of Sociology, 27*, 415–444. http://www.jstor.org/stable/2678628

Melucci, A. (1989). *Nomads of the present: Social movements and individual needs in contemporary society.* Temple University Press.

Næss, P. (2016). Built environment, causality and urban planning. *Planning Theory & Practice, 17*(1), 52–71. https://doi.org/10.1080/14649357.2015.1127994

Pesch, U. (2015). Tracing discursive space: Agency and change in sustainability transitions. *Technological Forecasting and Social Change, 90*, 379–388. https://doi.org/10.1016/j.techfore.2014.05.009

Rauschmayer, F., Bauler, T., & Schäpke, N. (2015). Towards a thick understanding of sustainability transitions—Linking transition management, capabilities and social practices. *Ecological Economics, 109*, 211–221. https://doi.org/10.1016/j.ecolecon.2014.11.018

Raworth, K. (2018). *Doughnut economics: Seven ways to think like a 21st-Century economist.* Cornerstone.

Saunders, C. (2007). Comparing environmental movement networks in periods of latency and visibility. *Graduate Journal of Social Science, 4*(1), 109–139.

Turner, R. A., & Wills, J. (2022). Downscaling doughnut economics for sustainability governance. *Current Opinion in Environmental Sustainability, 56*, 101180. https://doi.org/10.1016/j.cosust.2022.101180

XR Frome. 2019. https://www.instagram.com/p/B5afxlTHDaX/?utm_source=ig_web_copy_link&igshid=MzRlODBiNWFlZA== or https://twitter.com/ExtinctionF/header_photo.

Young, J. N., & Mayo, S. C. (1959). Manifest and latent participators in a rural community action program. *Social Forces, 38*(2), 140–145. https://doi.org/10.2307/2573934

Yu, K.-H. (2012). Formal organizations and identity groups in social movements. *Human Relations, 65*(6), 753–776.

CHAPTER 11

Government and Politics: Changes to Representation in Political Transition Arenas

Section 1: Government

Strengthening Ties between Tiers of Local Government through the Greening of Political Representation

The consolidation of a progressive and green political agenda in the town council. Since 2011, Frome Town Council (FTC) has been controlled by the independent group, the Independents for Frome (IfF). First winning 10 out of 17 seats in 2011, IfF has gone on to achieve 'full house wins' in the 2015, 2019 and 2022 local elections. As we will see, this has also been mirrored by gains by Frome-based Green party representatives in higher levels of government.

As we have seen in Chap. 4 and outlined in Table 4.1, IfF centres around a place-based approach where the role of the town council is a facilitator of broader community placemaking intentions. IfF also rejects party politics as a divisive 'poison' that infects the political immune system (Macfadyen, 2014; Burnett and Nunes, 2021). IfF's politics, and that cultivated by the political movement that sought to replicate Frome's approach (*Flatpack Democracy*, which we discuss below), can be broadly characterised as being within the 'progressive' branch of UK politics which centres around liberal social reformist and moderate socialist principles of egalitarianism, justice and environmental democracy (see Heine, 2012).

A. Burnett, *The Politics of Transition*, Palgrave Studies in Sub-National Governance,
https://doi.org/10.1007/978-3-031-40769-7_11

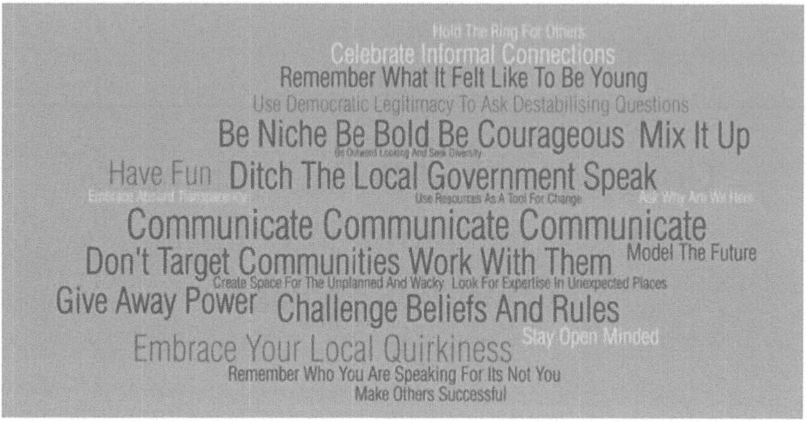

Fig. 11.1 Words of success: IfF word cloud, Breaking the Mould FTC conference April 2018 (IfF, 2018)

FTC has marketed its approach to political representation as one that is "breaking the mould" of party-political identities and its approach is to "infuse sustainability throughout", as purported through FTC's Breaking The Mould town council conference in April 2018. Some key phrasing used in this conference is demonstrated in Fig. 11.1, which illustrates the importance of informality, democracy, niche experimentation, community-led development, local autonomy and facilitating the empowerment of others to the town council's approach.

One FTC officer suggested that as a result of their governance, there was a more joined-up approach to achieving the town council's strategy. Peter Macfadyen (who co-founded IfF) and others were also involved in getting participatory budgeting off the ground in the town by working with political campaigning groups Public Square, My Society and DemoSoc.[1] These initiatives have since taken root and Frome's 'The People's Budget' has regularly asked local people to decide on projects important to them so long as they are aligned with FTC's strategic plan; local voters have a choice to select a first and a second preference project (for more discussion on participatory budgeting in enabling local democracy, see Manes-Rossi et al., 2023; Bartocci et al., 2022).

[1] See The People's Budget—Frome Town Council https://www.frometowncouncil.gov.uk/your-community/organisation-support/peoples-budget/

This has enabled FTC to transition from a 'traditional' town council to one that resources key functions to enact relocalised initiatives that operate within New Economy principles. For instance, employing a team of sustainability, resilience and participation officers and working in collaboration with community-based activists, such as Sustainable Frome, Edventure, Fair Frome Housing and the Frome Neighbourhood Network (FNN) (see Chap. 5). This has meant that the town council has managed to intensify and consolidate multiple sustainability initiatives in the town. A list of some of the sustainable schemes facilitated or supported by FTC is indicated in Fig. 11.2.

FTC's webpage has many other initiatives under the headings parks and open spaces, community projects, Keep Frome Clean, and health and well-being. Note the word 'sustainability'—which is open to interpretation (see Chap. 1)—has been avoided. In performing its enabling function, FTC recently agreed to write a letter of introduction to link the community energy group Frome Energy Coop (FRECo) to local landowners and explore opportunities to harness wind energy; as FRECo members suggested, "the support we are looking for from Frome Town Council helps us open the door and makes an introduction for us [to local landowners]" (though planning policy has generally not been widely

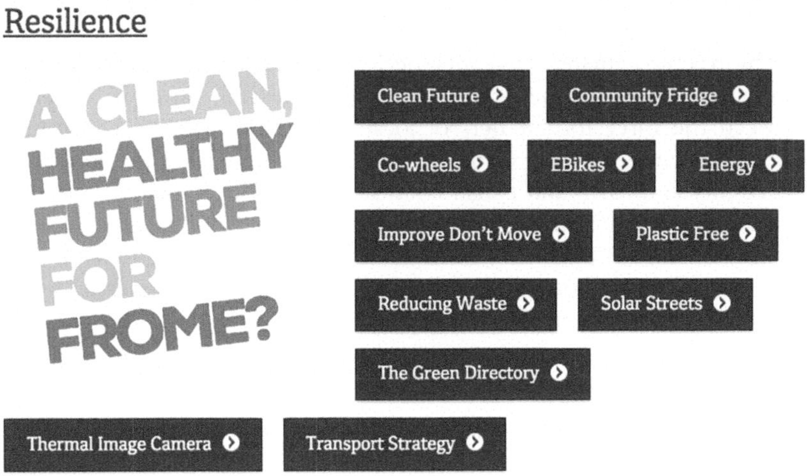

Fig. 11.2 Themes of Resilience—projects influenced or instigated by FTC (2019)

supportive of wind energy schemes due to changes from the 2015 Deregulation Act, see Rankl, 2023).

IfF's environmental action was perceived to be improving linkages with Somerset County Council (before it was abolished to make way for the new unitary council). An IfF councillor suggested the former county council was "absolutely delighted" by FTC's proactive green initiatives. Doing so supported improved local recycling, thus helping the county council meet its statutory duty to deliver on environmental issues under austerity localism (an issue that is particularly contentious in the wake of the new unitary council where a continued budgetary deficit has resulted in a 'financial emergency' declaration, which we turn to later).

The role of climate emergencies in stimulating environmental strategies and action. Along with many other town councils, FTC declared a climate emergency in April 2019 (the year that the 2008 Climate Change Act was amended to legislate a UK-wide commitment to achieve net-zero by 2050).

Locally, Peter Macfadyen was "busy working on a citizens assembly" when I spoke to him in 2019 and there was *A Case for a Citizens' Assembly in Frome on the Climate and Environmental Emergency* published in the same year.[2] These ideas were closely aligned to not only the proliferation of local 'declarations' but were given added countercultural flavour due to Macfadyen's ties to the environmental movement Extinction Rebellion (XR) and his role in cultivating democratic governance on climate action (which we discuss later in this chapter).

The role of climate emergency declarations has sparked intergovernmental alignment on the need to be carbon neutral, on paper at least, between the three former tiers of local government (Somerset County Council, Mendip District Council (MDC) and other district councils in the county and town and parish councils). At a county level, Somerset's Climate Emergency Strategy[3] was jointly developed by the five previous Somerset local authorities and is now taken forward by the new unitary Somerset Council. The development of collaborative strategies between

[2] See: https://www.frometowncouncil.gov.uk/wp-content/uploads/2019/07/Appendix-5.1-The-Case-for-a-Citizens-Assembly-in-Frome.pdf

[3] See: https://www.somerset.gov.uk/environment-and-food-safety/climate-and-ecological-emergency/somersets-climate-emergency-strategy/

tiers of local government is helping to enable a resilience agenda to become mainstreamed across local councils; for instance, a green eco business directory has been established (hosted by Somerset Council[4]), which aims to help residents make greener choices.

However, the extent to which this action through government forums is resulting in genuine 'transitions' in climate action is still to be determined, though as we will explore there is a greening of representation in new democratic spaces that open up opportunities to solicit multi-stakeholder environmental action. For instance, experiences in declaring and enacting climate emergency declarations have not been uniform.[5] Others have lamented the slow start to translate commitments into meaningful and urgent action by local and national governments. Also, a commitment is easy to make (and as we will see, in some ways so is winning power)—the challenge is sustained action in the face of political challenge.

There is a risk that declarations become a tool that deflects genuine action, especially in contexts that face (party-)political conflict on steering collective priorities. Hulme (2019) even goes so far as to suggest that such declarations are 'dangerous' as it might lead to a set of reductive indicators that force other concerns off the policy agenda and "forces other issues to be framed and subsumed within the politics of climate change" (p. 23). Meanwhile, there have been relatively few declarations that have highlighted the 'ecological' element of 'emergencies' (see Chap. 1). However, as we go on to discuss below, the mainstreaming of environmental discourse and the perceived need to act is helping to build the foundations of a broader movement to democratise climate action and decision-making.

"Shaking up" the Structure of Local Government

Challenging the planning regime? Based on its experiences of 'fighting planning' (see Chaps. 7 and 8) FTC has sought to try and empower the role of the town council to influence the tone of development in the town to be more proactive than reactive. While the council was able to restructure its internal governance to help empower others to influence

[4] See the Great Green Directory: https://www.somerset.gov.uk/environment-and-food-safety/climate-and-ecological-emergency/the-great-green-directory/ (NB: Frome's Resilience Manager also spoke of the need to develop something similar relating to supporting relocalised economies and development in the town back in 2014).

[5] For more discussion on the role of climate emergency declarations, see Ruiz-Campillo et al. (2021).

placemaking in Frome, FTC officers and IfF councillors found that the extant governance arrangements were not a real mechanism for significant influence on the planning regime. For instance, the FTC planning board had a function to comment on planning applications coming forward in the town, but this was felt to be too much of a reactionary role—rather than taking a more proactive approach to influencing local planning decisions and the type of development that should be encouraged in the town.

To challenge planning decisions further, IfF's second administration (2015–2019) sought to reconstitute the role of local planning governance. The planning board was reconfigured to become, as one FTC representative described, able to seek out "good development" and the "right" kind of developers to reflect Frome's cultural and historical character to enact innovative development models (see Chap. 8).

The election of Green MDC councillors (discussed below) also saw more formalised changes in expectations of greener development. For instance, in February 2021 Councillor Helen Kay put forward a motion to create Supplementary Planning Documents (SPDs) to set higher sustainability standards for developers and to create a Green Infrastructure Plan. However, MDC formally closed in April 2023, so this particular motion was not fully realised at district level.

Nevertheless there were improved relationships with the district council arising from the changing engagement patterns over strategic landholdings in the town (notably Saxonvale, see Chap. 8) led to what one FTC representative described as: "a different approach, [with] a different feel"; changes that have helped to structure a pathway for an alternative Mayday Saxonvale scheme to emerge in the context of the existential battle over Frome's identity, contested in the arena of urban planning, Frome's relocalised placemaking networks and the pursuit of viable solutions for development alternatives. (Though the Mayday victory at judicial review outlined in Chap. 8 in no way is a panacea for resolving contentious issues with planners and developers relating to securing solutions for appropriate development in Frome).

Changing relational ties in the context of unitary authorities. As we might recall, when the IfF came into power in 2011 there were political challenges between the newly independent-controlled town council and a Conservative-dominated district council. At various points, adversarial party-political tensions have both hampered and provided entry points to contest local visions for placemaking and development. Interviews with FTC and IfF respondents around 2015—from their perception at least—were clear that at the district level, MDC needed to transition away from decisions being made on political grounds (such as committee-based planning decisions), to be more objective, more understanding and respectful of Frome's needs and to move away from a position of conflict with the town council. (Though, as we shall see, IfF's protagonist attitude towards the district council was also a reason this conflict perpetuated).

As relational ties began to improve between FTC and the former district council (where the latter acted more positively to encourage a comprehensive solution for the Saxonvale site, see Chap. 8), a more radical shift in local government relationships was underway in the move towards unitary authorities. In May 2018, the leaders of Mendip, South Somerset, West Somerset, Sedgemoor District Council and Taunton Dean Borough Council issued a joint press release responding to a statement by Somerset County Council that it intended to sound out district councils and partners in the area about the appetite for a unitary authority to improve service delivery in Somerset. Various options were proposed, including an idea that Frome and Mendip could join Bath and North East Somerset Council (Frome is geographically close to the city of Bath and is used by many Frome residents to access employment, education, healthcare and other services). Also proposed at the time was that a new authority should be formed to incorporate Mendip and South Somerset as a possible, more representative, alternative (see Lee, 2020). The final decision was for a new unitary authority *Somerset Council* to incorporate the former Somerset district councils, which ceased to exist in April 2023. The new unitary authority now represents around 550,000 people, one of the largest unitary councils in England—Cornwall Council is currently the largest). In its logo (Fig. 11.3), the dragon head represents the "new beginning for local democracy in Somerset" referencing its depiction the county's historic coat of arms (Somerset Council, 2023).

Changes in local government configurations and political demography reflect broader political transitions that necessarily affect the flavour and

Somerset
Council

Fig. 11.3 Somerset Council logo (Somerset Council, 2023)

arenas of representation at a more local level (i.e., with these changes come reorientations between local representatives and the wider place-based networks they operate in). A major concern of some critics of the new council was that it would create further distance between decision-makers and local issues, and that what was required was more *localised* service delivery (discussed below).

Administrative changes can affect the ebb and flow of value alignment with political and placemaking networks. For instance, where and how new political identities take root (such as activating or demobilising certain values if party-political candidates fail to win seats in local elections), reflecting Connor's (2020) point that political values are socially reinforced rather than wholly predetermined.

For instance, there were concerns raised before the unitary authority was established (see Lee, 2020) that the creation of Somerset Council was politically motivated, whereby the "Liberal and Green- dominated Mendip area is swallowed up into the vast Tory heartlands that surround Taunton". (Somerset Council is actually currently controlled by the Liberal Democrats who have 61 out of 110 seats (BBC, 2022), so this particular concern that the Conservative party would dominate the new Council was not realised in its inaugural election).

At the same time, these administrative re-configurations can work in the favour of new political entrants. A Frome-based, Green Party councillor at Somerset Council, Martin Dimery, has suggested that proposals for the creation of a new constituency for Frome—which will come into force in the 2024 July general election—could potentially give entry points for a more balanced representation of party-politics. This new boundary would link to Radstock and Midsomer Norton to the north and Bruton to the south, which Dimery suggests might have the effect of splitting two different types of Conservative constituencies (one represented by the right-wing MP, Jacob Rees Mogg (North East Somerset) and the more 'progressive' constituency of Somerton and Frome, now represented by a Liberal Democrat, Sarah Dyke) (East Mendip Green Party, 2023; Lee, 2022). Dimery also suggests that this new boundary change may increase the chances of the constituency sending a Green MP to Parliament at the next election, surmising that it would become a "four-horse" race with the Green Party standing a good chance among the traditional political parties owing to its strong gains around Frome in local elections, which we discuss below.

The Politics of Being Authentically 'Independent' and the Greening of Political Representation

The "Green Wave" in and around Frome. The Green Party is a political party in England and Wales that considers itself as "a party of social and environmental justice". Its politics are based on 10 core values that emphasise democracy, equality, well-being, human happiness, protection of the natural world and non-violent conflict resolution. The party has traditionally been considered 'fringe'. It has been disadvantaged by the UK's First Past the Post electoral system, which favours more established parties and prevents smaller ones from gaining ground (which is arguably good for democracy).

The party has continued to make steady gains in recent years, reflecting the greater societal and political attention paid to environmental issues (such as the Paris Declaration, Greta Thunberg's youth activism and continued mainstreaming of environmental issues through climate and ecological emergency declarations). In 2015, the party celebrated a quadrupling of their membership since 2010 to 54,500 (Dennison, 2015) and it won an additional 99 seats in 2021 local elections. For instance, the party won 13 more seats—bringing the total number of Green councillors in the city to 24, leading the Greens to be the joint largest ruling party in Bristol City Council (Green World, 2021) and Bristol's Green Mayor was appointed in 2022 and championed climate assemblies for the city.

As of April 2024, the Association of Green Councillors now represents 746 Green Party councillors from 168 principal authority councils in England and Wales, of which 37 councils include Greens that are in power or part of the ruling administration. The party holds over 400 seats in parish and town councils (the level at which IfF and Flatpack Democracy have contended that there is no room for party politics) and—has—like IfF—begun to consolidate power in local government structures that represent Frome. However, in some cases, the Greens still have not yet translated a surge in its local party networks into votes (for instance, in the 2023 Medway unitary elections the Green Party put forward a candidate in every seat but failed to win any, and only came close to winning a seat in the ward of Cuxton, Halling and Riverside). Yet in the May 2024 local elections, the Greens gained 74 seats and the party now has 812 councillors elected; some of which are representing the party for the first time in places like Newcastle. On the back of this success, the party is planning to field a candidate in every seat in England and Wales for the anticipated 2024 general election.

In Somerset (notably stemming from Frome), there has been a steady rise of elected local green councillors, which has been described as a mini "green wave" (East Mendip Green Party, 2022).

- 2015: The Green Party successfully won three seats at the Mendip district elections for the first time (two in the ward of Frome Keyford and one in Frome Market ward).[6]
- 2017: Two Green Party representatives were elected to the former Somerset County Council, winning two out of three of the Frome-based wards (Martin Dimery—Frome East and John Clarke—Frome West), the first ever Green councillors at the Somerset county level.
- 2019: In the Mendip district elections the Greens won ten seats—eight of which were from the East Mendip (Frome) branch.
- 2022: In the Somerset Council unitary elections,[7] Frome-based Green candidates won four of the five seats won by the party (and out of a possible six seats)—one of which is a former Sustainable Frome director. The Liberal Democrat won Frome North along with the Conservative candidate.

However, in some cases these gains were on slim margins—the Green councillor Martin Dimery won only by four votes in 2017, with the Conservative candidate, Eve Berry (who also put herself forward in the 2022 Frome town council elections) close behind.

However, the Green vote share has continued to rise since then. In the first unitary election in 2022, Dimery stood in Frome West and won by over 30%. A by-election for the Westminster parliamentary seat for Somerton and Frome was held in July 2023 when the MP David Warburton stepped down following charges of political misconduct. In this by-election the Green Party candidate, Martin Dimery also put himself forward as a candidate and achieved 3944 votes, representing 10.2% of the turnout and a 5.1% swing to the Greens—the best ever by-election result achieved by the party (but still considerably behind the successful Liberal Democrat candidate, Sarah Dyke who won with 21,187 votes representing 54.6% of the vote share). The Green Party also did well in the Somerset

[6] See https://www.mendip.gov.uk/makeup (last accessed 7/6/19). There are also two councillors at Somerset County level—see https://www.somersetlive.co.uk/news/somerset-news/frome-home-first-green-party-46515 (last accessed 7/6/19).

[7] At the county level, wards are larger than at district level meaning the gains need to be larger to win a seat.

Council by-election in the constituency of Wincanton and Bruton (coming in third with 16% of the vote, above the Liberal Democrats and Labour).

At the unitary level, Dimery is Chair of the Environment and Climate Change Scrutiny committee, providing oversight for climate action in the council. The Mayor of Bruton, Ewan Jones, was Co-Chair of South East Somerset Local Community Network 2022—covering the areas of Castle Cary, Wincanton, Bruton and Milborne Port.

The continued election of Green local government representatives has helped to create the conditions for the consolidation of green, progressive political agendas in and around Frome and clearly there is a greater presence and oversight of environmental issues in the local and unitary arenas of representation. Working with the new 2015 cohort of Mendip district councillors created an anticipation of improved relations between the town and district council because of shared core green values. These new links between local government representatives were seen by IfF councillors and Sustainable Frome members as people they could "relate to", helping to bridge connections to Frome's active relocalised placemaking networks).

The increasing Green representation at the district level worked synergistically with IfF's continued connection to Sustainable Frome. In 2019, a former Sustainable Frome director, Sheila Gore, was elected as an IfF councillor and was the Mayor of Frome, while at the district level one of the elected Green Mendip district councillors was also a former director of Sustainable Frome. Gore and others actively worked on consolidating the connection with these new Green politicians who were operating at the interface of the lower tiers of government to champion Frome-led politics. This resulted in a Frome Councillors Forum, where town, district and county councillors met to jointly work on issues that affected the town (see East Mendip Green Party, 2018) (again echoing the importance of political representatives 'understanding Frome' in facilitating effective ties to support placemaking that reflects local interests, see Chap. 6).

The politics of green and 'independent' values. To overcome the inherent challenges of the UK's first-past-the-post electoral system,[8] Compass—the leftist political democratic movement—has been active in calling for greater tactical voting, depicted in its Progressive Alliance model. Here, (typically centre-left-leaning) voters agree to vote for the

[8] See: https://www.makevotesmatter.org.uk/first-past-the-post

'next best option' if it looks like their preferred party won't win in their constituency (an approach that echoes the tacit agreement between IfF and the Green Party not to compete with each other in Frome's town council local election). This could well be one strategy that can encourage value alignment at a national level, although as Dimery argues in one of his many positioning papers on local politics, if people want to vote tactically they will, without the need of a campaign to encourage them to do so (Green World, 2023); indeed his supporters said they were switching their vote from Green to Lib Dem to stave off the Conservatives (see Walker, 2023).

However, voting tactically to consolidate a green agenda challenges the extent of political 'independence'. For instance, in the 2015 local elections there was a (tacit) agreement between IfF and the Green Party that the Greens would not stand at the town council level and that IfF would not seek to stand at the district level—which one IfF councillor described as "a kind of rational decision-making from both groups"—indicating a steer on shared political strategies. An IfF councillor was keen to express that it wasn't a deliberate strategy, but this approach called into question the extent of 'independence' owing to a 'green' value alignment within local political arenas. Or in this councillor's words, if the Greens had stood against IfF locally in some wards, it could have meant IfF lost seats, in effect "coming from nowhere, they could pinch enough votes off us to do real damage … [If the] Green Party managed to shaft Frome's greenness [it] would be insane".

The limits of independence in steering political transitions was recognised by an IfF councillor who declared that "I've got to work out a way of supporting the Greens without erm, confusing too many people about my own independence [laughs]". However, at the 'Do Democracy Day—taking back power and politics' event I attended in September 2015 at Frome's Cheese and Grain,[9] one participant strongly questioned why IfF councillors were making tacit alliances with the Greens, a response which was shelved for discussion outside of the event (which was being recorded by the social movement Unlock Democracy).

Moreover, one respondent was surprised when I pointed out that not all IfF voters might vote Green, indicating an implicit value alignment that may not always be more widely shared amongst voters. This illustrates a

[9] See https://www.frometimes.co.uk/2015/08/25/unlock-democracy-political-workshop/ A 'short version' of the day was available at https://vimeo.com/140496051 (last accessed 17/1/19).

clear tension where value alignment can either be negatively construed as a political strategy by an opponent (i.e., when different values lead to othering), or positive synergies for mutual co-benefit (i.e., if values are shared)—highlighting the politics of steering for transitions. That is, if the shoe was on the other foot and a cluster of people influencing local politics in this way had, say, far right-wing orientations then the capture of Frome's political architecture would be construed negatively, rather than a positive condition to enable relocalised intent. It is also worth noting that locally, as indicated in the quote above, IfF are sometimes referred to as the "IfFies", implying a degree of suspicion or political othering—particularly by party-political opponents.

'Party politics' as a dirty word and the co-option of 'independence'. The politics of discursively marketing oneself as independent was also called into question over the controversial instance of party-political actors claiming to draw on the 'independentness' of their politics. For instance, Neil Burns, the Labour candidate in the 2018 Frome Market Ward by-election commented on the Liberal Democrat election flyer and suggested this "had the 'look and feel' of an IfFy flier, and whose only mention of Liberal Democrats was hidden away in 6 point type" (Somerset Labour, 2018).

The 2019 election leaflet of a Liberal Democrat, Drew Gardner, was also made to look like the black-and-white design used by IfF in its previous campaigning alongside the words "independent-minded", despite being a member of a party-political organisation. This drew a degree of local ridicule in the local newspaper and suggests how 'independent' values and positionality could be co-opted by some party - political actors to claim the space that independent politics might cultivate.

In the 2022 elections, a similar phenomenon took place with the Conservative candidate. Eve Berry had previously been elected to the town council as a Conservative party candidate and stood as a Conservative in the 2019 Mendip district elections and the 2022 Somerset Council elections, yet she stood as an 'independent' in the 2022 local elections in Frome.

Clearly, IfF has consolidated their campaign for independent politics in the town when party-political candidates fear rebuke for showing their political true colours and party-political candidates resort to chameleon-like behaviour (or as the Labour candidate Neil Burns argued, "[It] [c]omes to something when you think your best chance of being elected its to pretend to be someone else" (Somerset Labour, 2018)).

However, this has most definitely not beaten off party-political identities for good since these are quick to reassert themselves higher up the political value chain, i.e., above town council level electoral ballot papers. Thus, it is important to treat the discourse of independence with some caution, particularly when used by party political actors as a discursive tool to seek election.

Last group standing—IfF: a democratic victim of its own success? Moreover, the effect of party-political independence is causing some unusual effects on democratising local politics in Frome. As I previously noted, the fact that in the 2022 FTC local elections in all but one ward (where Eve Berry put herself forward as an independent candidate) none of the other wards were contested, IfF councillors now find themselves in a situation where they are co-opted into the town council without needing to be elected, which is a strange inversion of its wider democratic mandate.

Relocalised Politics in the Context of Unitary Governance

Geography, distance and the politics of representation. Somerset County Council declared that the abolition of district councils would improve efficiency and enable more integrated services on waste, parking, planning and highways. Yet a joint statement by the district councils questioned this and argued that the claim that a (centralised) unitary authority is required to protect local services was inadequate.

In particular, an article in the local paper *The Leveller* in 2020 (Lee, 2020) expressed concerns raised by district councillors about the distance of Frome (56 miles) from the county town of Somerset, Taunton (where the Somerset Council offices are based). Concerns were raised that the unitary council would accentuate the lack of local understanding of Frome-based issues in a far greater way than was even the case with Mendip District Council (MDC was based at Shepton Mallet, 12 miles from Frome). Concerns around the size of the geographic area of Somerset county were also flagged as an issue, which has a population of over half a million residents, again being at odds with the more localised representation that cultivates understanding of a given place.

Bridging the gaps: expanding the reach and depth of localised transition arenas. In the transition to unitary governance, Somerset Council is working with town and parish councils and other important public and local stakeholders via 18 new Local Community Networks (LCNs) to bridge the gap in representation as a result of the abolition of the district tier. Some services are being managed on a regional hub basis—suggesting a more decentralised hubs and spoke approach that could tie in with wider governance arrangements of citizen assemblies and other community-based networks, if structured and resourced effectively. (And as we saw above, these have become arenas for a further greening of local democratic spaces, as in the case of a Green Party representative being a co-Chair of a Local Community Network).

These more localised spaces for representation could also work in tandem with community-based initiatives such as the Frome Neighbourhood Network (FNN) and its adopted Transition Streets model (see Chap. 9) and help to bridge these entities into more constructive and continued engagement in community-led planning arenas, such as the planned revision of Frome's Neighbourhood Plan (see Chap. 10).

Concerns over the centralising effects of unitary governance also reflects the concerns raised by FTC staff and IfF councillors of the need for a further devolution of powers through and extending beyond the provisions of the Localism Act, including on planning decisions, which many I spoke to thought should be delegated to the town council level, more akin to municipalism in European countries. (See for instance Comer-Schwartz et al., 2023 for Labour's suggested 'Community Power' approach).

Financial black holes in local authority budgets: the elephant in the room of a functioning democracy? However, a major sticking point in making new local democratic arenas, such as LCNs, work under the new unitary governance umbrella is the massive budgetary black hole facing Somerset Council. It has only been in existence since April 2023 and as of December 2023, the council declared a 'financial emergency' owing to a funding gap of what now stands as £100m for FY2024-5, driven by, amid other factors, the rising cost of adult and social care. As a result, the new Council may have to sell some of its assets and it plans to increase council tax rates by 5% and implement cuts to local services and axe around 1000 jobs at the authority. Notably, it also plans to sell assets including its landholding on Saxonvale (see Chap. 8).

The financial situation of other local authorities looks dire. 2023 also saw some major local governments, including Birmingham City Council, issuing a Section 114 notice (see Fig. 11.4), which is effectively a declaration of bankruptcy where costs are likely to exceed a local authority's income for a particular financial year, which is prohibited by the Local Government Finance Act 1988. In these cases, a council cannot commit to any new spending other than for essential services. Other local authorities plan to follow suit in the near future and many others are introducing additional curbs on their spending.

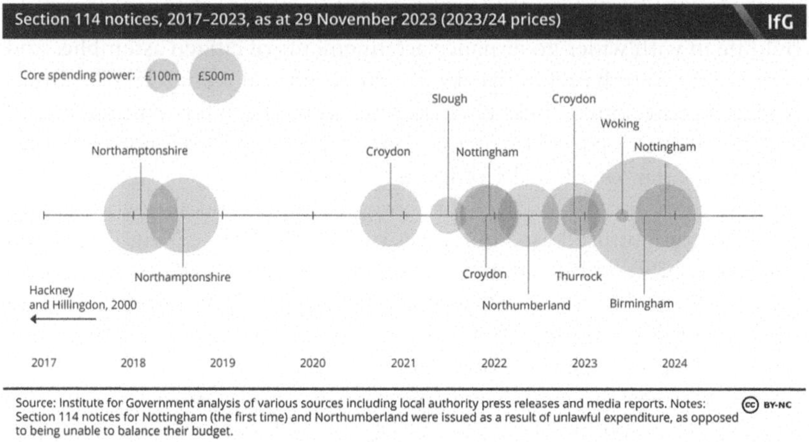

Fig. 11.4 Institute for Government analysis of various sources including local authority press releases and media reports, Section 114 notices by local authorities (IFG, 2023)
Note: Section 114 notices for Nottingham (the first time) and Northumberland were issued as a result of unlawful expenditure, as opposed to being unable to balance their budget

The serious and significant financial challenges facing local government, compounded by long-term austerity measures and the more recent cost-of-living crisis are a risk to broadening out democratic reform. National government rules on reallocating spending from capital to revenue budgets may affect the flexibility of plugging these financial gaps (although action is being taken by the Department for Levelling Up, Housing and Communities (DLUHC) to help support the cash-strapped local governments).

On the other hand, in response to the situation of financially crippled local authorities this could yet provide opportunities for a further entrenching of localised solutions—incubated through the localism agenda and the wider climate and ecological emergency response—where local councils at the parish, town and city scale act in tandem to democratise spaces for climate and civic action (see Chap. 10). For instance, those championed by the Flatpack Democracy and XR movements or the resilience initiatives facilitated by FTC (see Fig 11.2). However, as we have seen throughout this book, resourcing this and creating the conditions to mobilise change is far from straightforward.

Section 2: (Flatpack) Politics

Flatpack Democracy—Scaling Up and Out

Igniting interest in indy politics—blazing a trail with the Flatpack approach. Due to the experimentation with independent politics and its potential to engender more sustainable and inclusive politics, IfF has stimulated a desire for "revolutions" in other towns and cities, in the UK and abroad. FTC and IfF were deemed by one FTC representative to be "blazing a trail" in terms of their approach and encouraging others to self-replicate 'indy'-led councils.

At a national level, the model—based on Frome's experiences—has been championed by the founder of Sustainable Frome and co-initiator of IfF, Peter Macfadyen and supporters of the Flatpack agenda. As part of this wider movement, targeted support has been offered to candidates seeking election as an independent councillor by mentors (current and former IfF councillors or independents from other UK local councils) or to share insights and create connections with others seeking to change within local government under the Flatpack model.[10] Macfadyen has also forged ties with similar left (and green) radical political groups in Europe, such as the Alternativet Party in Denmark, the Pirate Party in Iceland and Podemos in Spain, as well as in the US.

Macfadyen's two books '*Flatpack Democracy—a DIY Guide to Creating Independent Politics*' (2014) and '*Flatpack Democracy 2.0—Power Tools for Reclaiming Local Politics*' (2019a) have been key in influencing this approach as an artefact to share Frome's model of independent politics, which have sold over 10,000 at the time of publication of this book. Macfadyen actively posted in Flatpack Democracy blog (which is no longer available, and is now directed to a single page on the books' publisher's website).

In the 2015 local elections, a national independent political actor suggested they were buying in bulk copies of Flatpack Democracy and "distributing them as fast as we could" to those who could be interested in forging more 'independent' politics across the country. At this time, there was also an idea to hold a "regional conference of revolutionaries". By 2019, there

[10] A website IndieTown http://www.indie-town.uk/ had loads of helpful information on the spread of the Flatpack Democracy movement that was updated by a Frome-based individual, in collaboration with Macfadyen but it is, unfortunately, no longer available (last accessed 26/1/19).

Fig. 11.5 The Rights of Spring—fluttering of 'indy' councils, May 2019 (Macfadyen, 2019b)

was a strong swell of interest from other towns in the South West as places that were, according to Macfadyen, "definitely going for revolutions", encouraged by Flatpack Democracy—as illustrated in Fig. 11.5.

The movement for indy politics became more formalised in 2021 in a Flatpack Democracy national campaign, which together with the two books helped to share the experiences to replicate Frome's 'Do It Yourself' political model. In the rest of this chapter, we discuss the insights from the 2021 campaign, why the translation of interest to indy-led revolutions has not been a quick win and what the co-option of Flatpackery into XR's democratic agenda means for the future of progressive and green politics in the UK and beyond.

The Flatpack Democracy 2021 campaign. The efforts to propel independent politics as a mechanism to engender locally focused politics were accelerated by the national-level Flatpack 2021 campaign. It described itself as "a decentralised campaign to support thousands of non-party community councillors to win seats in council elections in May 2021, directing them to the resources/information to run a successful campaign.

Its goal was to see local councils flooded with independents through their targeted support and mentorship programme based around Frome's experiences and Macfadyen's two books (see Fig. 11.6) and the campaign team organised various online events[11] (I attended some of these). There was also an associated website for the campaign (which has now been

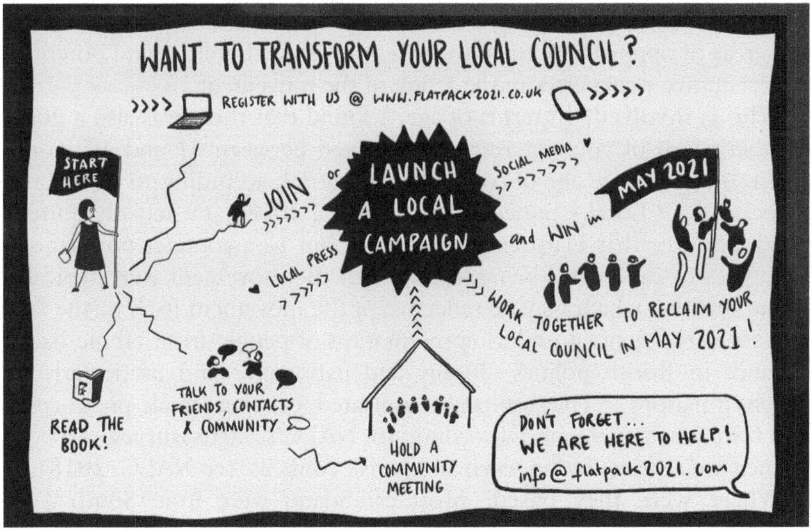

Fig. 11.6 Flatpack Democracy 2021 Campaign Team 'how-to' on launching a local independent campaign (Flatpack Democracy Campaign Team, 2021)

[11] See: https://www.flatpackdemocracy.co.uk/flatpack2021-off-to-a-flying-start/ (last accessed 10/7/22, this website is no longer live) and https://www.demsoc.org/public-square/case-studies/independents-for-frome-a-different-kind-of-local-politics (last accessed 12/6/23).

taken down). The campaign team wrote about its desire for community-led politics, run by 'real' (as opposed to party-political) people:

> *We believe councils should be run by local communities, for local communities. We want to help people reclaim their councils from party politicians and put real people back in control. We're aiming to catalyse a mass movement of independent councillors whose sole aim is to carry out the wishes and wills of their local community. Our big goal is that every council election in May 2021 is won by an independent community-led group.*[12]

As part of follow-on work in this 10-year longitudinal study on sustainability transitions, planning and independent politics, I engaged with the Flatpack 2021 campaign team to find out more about it. I wanted to find out more about who was involved (e.g., those standing as a councillor or willing to support candidates), people's knowledge and experience of Flatpack, the governance of their group, links between indy politics and climate emergencies, ties to environmental groups and their party-political membership, if any. In doing so, I was hoping to uncover possible trends and areas of opportunity to cultivate less adversarial politics and potential representative weaknesses in the fabric of the movement.

Who is involved? In terms of age, I found that there was also a good representation of younger respondents aged between 31 and 40 (compared to the average age of councillors being 59; according to a National Association of Local Councils study (NALC, 2018), it's actually quite a good indicator that Flatpack was reaching out to a younger population. The types of candidates being attracted to the movement were typically White British—which may be reflective of the movement itself or the fact that there is a general under-representation of people from ethnic backgrounds in British politics—locally and nationally—and particularly in devolved nations.[13] The results also indicated a strong female presence in the Flatpack movement (according to NALC's 2018 survey, 40% of women are represented in town and parish councils; see NALC, 2018).

Where were they based? Most candidates were from South-West England, reflecting geographic ties and spatial proximity to Frome in Somerset. Out of those I engaged with, there was significantly less representation in other regions of England (two from Wales and none from

[12] https://flatpack2021.co.uk/about-flatpack-2021/ (last accessed 10/10/21)—NB this website has now been taken down.

[13] See: https://lgiu.org/publication/diversity-in-political-representation-in-the-uk/

other areas of the UK). This suggested that in 2021 Flatpack hadn't yet started to make much of a dent in other parts of the UK outside of England.

What is the 'political' make-up of independent councillors? Existing party-political persuasions in the 2021 cohort indicated that while some had been an episodic member of a political party, they typically had not been committed to political parties—suggesting that independent politics may struggle to convert those with longstanding and more entrenched party-political views. While there was some representation of those that had previously voted Conservative, in the main, the 2021 cohort didn't, and many had voted Labour/Cooperative Party; even more common was for them to have voted Green, echoing the strong alignment between IfF and the Greens at local and district (and now unitary) elections discussed earlier in this chapter.

What experience did they have of being in political roles? Over half had been an elected councillor at various government levels before, and most were current town councillors. This indicates that there is a solid amount of political capital that can navigate (and possibly circumvent) the political system within the emerging indy political movement, if channelled effectively.

How easy was it for them to get elected? Of those that were elected as an independent, notably those at the town council level, respondents generally felt they were elected with at least a comfortable majority, if not way ahead of party-political candidates. This suggests that where independents do get elected, they are 'pushing at an open door'—either through the attractiveness of the candidate or through the offer of a new 'beyond party politics' model. Though, this would require further research to unpack this more.

What motivated them to get involved? In terms of motivations for getting involved in the Flatpack movement, mainly this was driven by a belief that their council needed "shaking up". This indicates a challenge of using existing councils to deliver on localism as there is clearly a perception of ineffective capacity within local councils to deliver effective representation at a local level. The second most common reason was those involved saw Flatpack as a genuine alternative to (divisive) party politics. Most were encouraged to get involved after reading Macfadyen's first book (Macfadyen, 2014), and a lesser amount his second (Macfadyen, 2019b). In some cases, there was a translation effect of candidates putting themselves forward which also encouraged a friend or colleague to read one of Macfadyen's books.

How was Flatpack encouraging experimentation with different governance approaches in local politics? Overall, the most important governance tool used by elected councillors was the creation of Ways of Working (WoW) to establish core principles and values in the group (see Chap. 6). However, no councillor (or group of councillors) had yet to formalise these in their Standing Orders as FTC had done in 2015. Keeping meetings and governance light-touch by meeting in more informal spaces was also considered to create effective conditions to open and improve Flatpack governance (see Macfadyen, 2023).

To a lesser extent, engaging in more reflective governance was also being fostered by following the Flatpack model. For instance, monitoring and evaluation is a well-established principle of effective project management and if embedded within more reflexive governance it helps lessons to be learnt and improve the implementation of projects (see for instance, Lodder et al., 2020). However, there was scope to improve monitoring and evaluation at a local level of Flatpack-inspired outcomes (and local council actions more broadly); while some reported to be monitoring and evaluating their work, others didn't know if they had a monitoring and evaluation process in place, or were doing so but in a less systematic way.

Are there strong links between Flatpack candidates, well-being and (local) environmental organisations? Councillors tended to be members of a community-based or interest group, which in some cases included being involved in environmental groups, including XR, Transition Towns, other local environmental groups or social enterprises but this is certainly not the case for all councillors/candidates. The links between independent politics to environmental and well-being issues were not always particularly strong. This suggests that if replicating the Flatpack model, Frome's mix of strongly sustainable and relocalised (*Transition/* New Economy) influenced initiatives might not be emulated or the necessary conditions discussed in earlier chapters of this book may not be in place to achieve the same strongly resilience-focused outcomes (unless perhaps they have similar strong 'environmental' place-based histories). While climate emergency declarations had typically not been declared at parish council level those made at higher tiers of local government had been influenced by some indy councillors, though this certainly wasn't the case for all of them.

What did the Flatpack 2021 Campaign Team do well to help the movement scale up and out? The efforts of the Flatpack Campaign team were widely acknowledged and appreciated in the way they cultivated a supportive environment. The campaign team was felt to have been effective at organising meetings via online forums and engagement tools (such as private Facebook and WhatsApp groups as well as regular Zoom meetings). Through the use of these tools, the provision of support and information and the Flatpack 2021 mentorship scheme, respondents described this type of support as "vital" and "inspiring" which linked like-minded people in a "supportive and welcoming atmosphere".

Their support demonstrates a prominent transmission effect that stimulated knowledge sharing and role transitions based on the ideas of Flatpack Democracy. As one respondent suggested:

> *I think the events Flatpack organizes [sic] -, the books, the emails, the WhatsApp groups, [and] the Thursday evening catchups—are all part of creating an inclusive environment where people can get the courage to try to make the change they desire to see.*

What suggestions did candidates have to improve Flatpack's reach? In terms of what could be done better to improve the movement's scaling strategy and support for its members, some candidates supported by the campaign team were satisfied with its existing approach. As one Flatpack candidate suggested:

> *Keep doing what you're doing—you lit a fire under us here and we're glad we got involved—we would NOT have without your support.*

There were some suggestions by candidates to improve the coordination of the Flatpack campaign and its strategy to scale out the indy political approach and to amplify its potential to effect representative and inclusive change. For instance, continue to reach out to women, make more targeted efforts to engage people of different ethnicities, improve social media campaigning by knowing their audience for different platforms, hold regional campaigns and be more targeted at higher levels of government to communicate the benefits of independent politics beyond parish, town and district councils.

So, What Is the Effect of Flatpack on Local-Level Political Transformation?

Scaling up indy revolutions. Engagement with the 2021 Campaign team suggests that indy politics was reaching a younger and female audience—which is promising as a political attraction that transcends the traditional boundaries of older, male representatives in our political systems. That many involved had previous experience working in political systems (i.e., as councillors at different tiers) provides useful 'insider knowledge' of how to navigate the system by working in partnership with those who haven't; this can give rise to emergent, fresh perspectives of what change might look like in local government (as with the blend of Macfadyen's and Usher's countercultural and regime insights in winning and consolidating IfF in the town council in Frome—see Chap. 5).

The unsteady steps to scaling up and out. Some councillors and candidates involved in the 2021 campaign felt that the Flatpack model was delivering everything they thought it would, which is very promising as a potential model and package of 'power tools' for local, radical democracy (see Macfadyen, 2019b). However, there were issues in replicating this; some respondents struggled somewhat to carry out the basic ways of organising suggested in the two books, highlighting that not all contexts and personalities will carry the Flatpack movement into the fabric of local government uniformly. For instance, when reflecting on the capacity of IfF to scale up, the emergent methods of *Transition*, such as Open Space and World Café (which have also been used by IfF and Frome's Neighbourhood Planning consultations), were deemed replicable to solicit new approaches to placemaking.

However, an IfF councillor I spoke to was wary of how these groups would seek to replicate Frome's experiences in different contexts. For instance, echoing how Transition Initiatives might try and follow the steps to transition proposed by the Transition Network verbatim, without reflecting on how these can unfold on a more emergent basis (rather than following linear, structured 'steps'). In fact, as we will see below, the illusion that communities can follow 'steps' to unlock democracy is potentially naive and ignores the realities, and complexities, of local politics (Fig. 11.7).

Brokerage between government layers and the challenge of breaking through. Flatpack has made gains at local councils, but less so higher up government tiers. To overcome the challenge of breaking through

Fig. 11.7 An example of an independent campaign in East Devon—though not all independents necessarily prescribe or were influenced by the Flatpack Democracy movement (Image taken by the author, 2023)

political regimes (particularly higher up the value chain) one national independent political actor suggested that independent councillors could form a 'party' and retain their autonomy to enable leadership at a national level. However, when I spoke to Macfadyen to catch up on the status of the movement even he still wasn't sure about whether the Flatpack model, as a place-based political movement, could ever work effectively and be replicated higher up the political value chain (i.e., district, unitary and national levels). Speaking in 2015, some interviewees I spoke to felt that the lack of networks from the local level up was perceived as a blockage for independent councillors to incite scale-shift.

It is particularly noteworthy that the 2022 Somerset Council elections saw IfF candidates (unsuccessfully) put themselves forward into the spaces of higher-tier representation. However, the extent to which there is an appetite for IfF to battle for a political presence beyond the town council remains to be seen, particularly amid the difficulties faced by 'fringe' parties to break through the established party political scene. (As we have discussed, differences between voting practices at national or lower tiers of government can sometimes mean that support for the Greens is more localised—that is, people are more likely to vote Green in local elections rather than national—which could affect potential coalitions between independent groups and Green councillors at higher tiered local government level.)

Severing ties to the Flatpack Democracy movement. Late in 2023, I checked in with Peter Macfadyen to get a final update on how Flatpack was challenging party politics and encouraging political innovation and alternatives. It's not what I was expecting, having believed in the promise of Flatpack to fundamentally alter divisive political systems. Despite gains in independent councillors in 2023, particularly in South East England (Wright, 2023), Macfadyen had effectively called it a day with the Flatpack Democracy 'movement' to which he had dedicated so much of his energy to. The blog that Macfadyen had so actively maintained was no longer available. There were no more plans for campaigns to put up indy candidates like in 2021. It felt very much like the umbilical cord to a nascent movement had been cut.

He directed me to the article 'Flatpack Democracy 10 years on, was it either?' he wrote in the summer 2023 edition of the New Economy magazine *Stir to Action*. In this sobering read, Macfadyen reflects honestly on the gains and limitations of the Flatpack approach and the limits to scaling up. Here, he writes about the impossible task of translating Frome's experiences into "Ikea-like instructions to unpack the holy grail of democracy" (Macfadyen, 2023, p. 9). On a positive note, he acknowledged that IfF managed to open up democracy to the people of Frome and encourage people to get more engaged in topics discussed and enabled by the town council. Yet he signals that the adversarial, protagonist attitude (described in earlier chapters)—i.e., niche resistance—cannot be sustained for long. Or as he suggested, "It turns out that a gung-ho 'ready, aim, fire' approach is hard to maintain and certainly hard to replicate" (Macfadyen, 2023, p. 9). Macfadyen laments, as we have discussed in this book, that even democratic innovation and experimentation revert to regime-like states:

"There is a depressing pullback to the norm in many aspects of council activity...the more interesting ideas get lost in bureaucracy, and before you can blink the old men are back running the show" (Macfadyen, 2023, p. 9). He cites the case of Bexhill-on-Sea in East Sussex as an example, where "winning control of a council is the easy bit and without constant work and vigilance the zombie returns" (Macfadyen, 2023, p. 10).

For Macfadyen, nothing has really changed in the "arcane and dysfunctional" local government systems Macfadyen and others sought to challenge through IfF back in 2011, despite the promise of localism (see Chap. 8). Macfadyen argues that far from empowering the 'man on the street', localism empowered other tiers of government higher up the chain, such as district councils. Though he cites the exception of the General Power of Competence (see LGA, 2013) which enabled Frome's town council to have autonomy on its spending decisions. Using this power, FTC obtained important assets for the town, such as the new council buildings and a stake in the Saxonvale landholdings, see Chap. 8. These have been essential for Frome's relocalised agenda to have a genuine chance in forging development alternatives and incubating ideas and solutions for placemaking in the town.

However, he is brutally honest in the limits of his efforts to mobilise a large-scale democratic structural shift through infiltrating the politics of local councils: "I have failed in catalysing large-scale change, or supporting the groups that have gained power in a way that has led to radical change" (Macfadyen, 2023, p. 10). Rather, he feels he has fostered a support network for groups of people who already understand that local councils need to work with communities. Macfadyen still continues to support Flatpack groups to help them in their electoral campaigns but this is now more on a unilateral basis rather than seeking to build a movement. He also supports local councils through training on creating resilient, cohesive and green communities alongside the former co-chair of the Transition Network, Peter Lefort.[14]

Instead of channelling his energies into the 'stand-alone' Flatpack movement, he has decided to put his countercultural energies into pastures new. Like with the experiences of the ebb and flow of Sustainable Frome and its spin-off groups, Flatpackery has itself been transformed and subsumed into new radical agendas of the New Economy that build

[14] See: https://blackmountainscollege.uk/events/parish-and-town-councillors-playing-a-key-role-in-creating-green-and-resilient-communities/

parallel democratic systems "to rebuild governance from the bottom up" (Macfadyen, 2023, p. 10) in the context of consolidating participatory climate governance (discussed below). Nevertheless, Flatpack Democracy continues to inspire budding independent councillors to stand, such as in Portland Dorset where a group of independents recently took control of the local council and won two seats at the unitary level. For these individuals, the promise of do-it-yourself democracy is still very much alive.

Pastures new: building a politics of hope through participatory and grassroots democracy. Engagement with the Flatpack Democracy 2021 team and my wider research revealed there was a perceived misalignment between tiers of government to develop effective place-based solutions to address climate and ecological emergencies. This challenge adds an additional layer of complexity to broker collaborative relationships within local government.

To address this, XR has become increasingly present on the left-leaning democratic scene to ignite climate action from the bottom up.[15] XR has suggested that reforms (e.g., proportional representation) and democratic models (e.g., climate assemblies) can help to prefigure, or actively model and construct, an end to the first-past-the-post system in favour of something more decentralised and representative of the will of 'the people'. Macfadyen suggests that other, complementary political groups, such as Cooperation Hull (which is seeking to build a People's Assembly), also hold much promise to realise the Flatpack vision, but in a slightly different way. (And indeed its origins has links to the "tentacles" of Flatpack, as Macfadyen suggested).

Flatpackery is also one of the many different democratic solutions endorsed by XR to building self-governing climate solutions based on radical inclusivity, active listening and trust into the 'DNA' of political systems; although as Fig. 11.8 depicts, within the potential solutions to democratic governance suggested by XR, Flatpack is also seen as one of the more 'complex'—lending itself to Macfadyen's conclusion (and that of this book and other sustainability transitions scholars) that step-by-step models to seize power are hard to replicate.

[15] Here its approach is not to work with any one political party, but to cultivate a partnership between entities such as Compass, XR and new left progressive movements and parties, such as Transform which aims to build a new party of the left and merges other entities such as the Breakthrough Party, Left Unity, the Liverpool Community Independents and the People's Alliance of the Left) (Salvage, 2023).

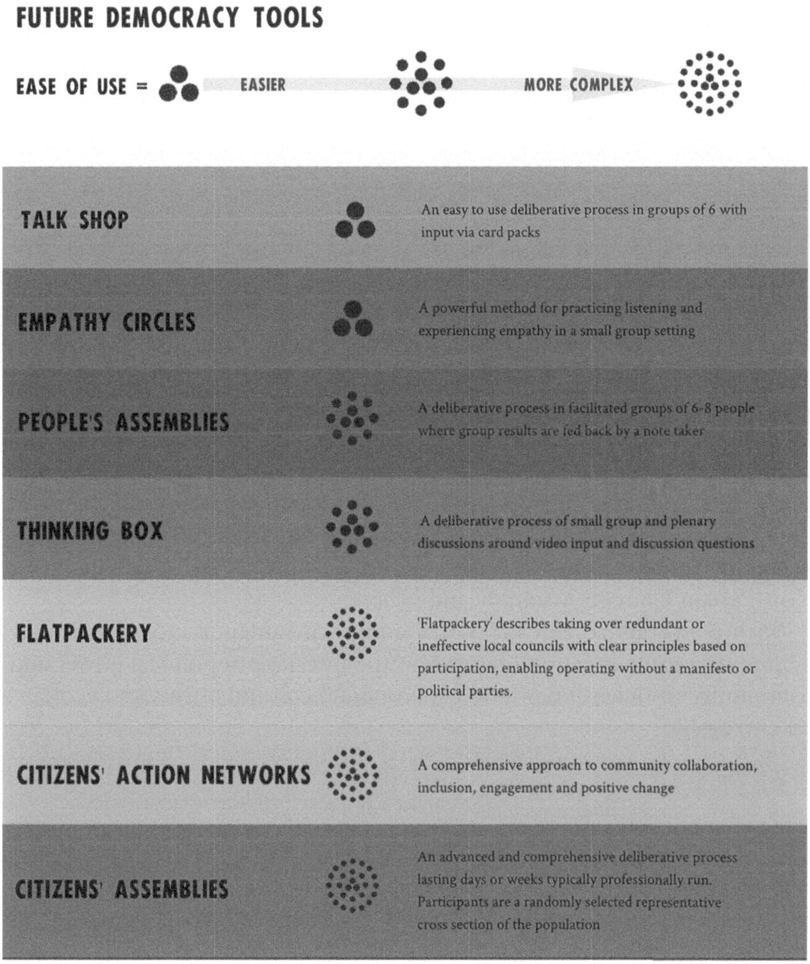

Fig. 11.8 Flatpackery and the various forms of self-organising democratic systems for radical climate action (XR Democracy Hub, 2023)

Beyond XR, the *Humanity Project* is a 'movement of movements' that reflects a consolidation of the facets of the New Economy woven into a diverse and shared vision of building more humane democratic systems. The project aims to support people with the tools to re-imagine democracy from the grassroots up and its vision has been co-created with artists and activists converge seeking to realise:

> *A new culture of inclusive democracy, crafting a new common sense, delivering the solutions we need... We are upgrading our democracy to a system that works for everyone, replacing the broken operating system of politics-as-usual.*[16]

In the words of the Humanity Project's website, it aims "to support the emergence of this new ecosystem; to reimagine democracy and public discourse for times of polycrisis". Its goal is to establish a permanent House of Citizens and "a new constitution of sortition-based citizens assemblies, setting in motion a new democracy culture around the country".

As well as a core team seeking the buy-in of communities at scale, it is a fusion of multiple COVID-19 community support systems, the architecture of XR's democratic ideas (the founders of XR also co-found the Humanity Project). It is informed by the needs arising from the cost-of-living crisis and culturally diverse.[17] The movement includes and draws on the experiences of other movements, such as The Cost of Living Alliance, Trust the People, Cornwall Climate Action Coalition, Flatpack Democracy, Nottingham Climate Assembly and others.

Such convergence between social and environmental movements may help to overcome structural barriers at the interface of political power and community to unleash new and experimental cultural norms and ecological citizenship.

Conclusion

This chapter has examined the changing relationships between local government tiers which are increasingly mediated by climate and environmental discourses. For instance, we are seeing (shared) implementation of various sustainability initiatives and the harmonisation of discourses through climate emergency declarations at different scales (brokered by an increasingly active XR movement and its affiliates).

This is in tandem with changing administrative boundaries, and with that, new democratic entities being established. For instance, there has been a distinct shift in the underpinning architecture of local government,

[16] See: https://humanityproject.uk/the-plan/

[17] See https://humanityproject.uk/

particularly in Somerset but also in other places in England in the move towards unitary councils. Localised democratic arenas are offering modes to expand arenas for community input in decision-making (e.g., bridging local-unitary council representation gaps, such as LCNs, participatory budgeting or citizen and climate assemblies).

There are signs of gains by green and progressive factions, such as an increasing number of Frome-based and green councillors elected to the higher tiers of government since the 2015 local elections (which invited in tacit strategic agreements not to split a more progressive vote). At the same time, IfF and the Flatpack Democracy movement have been cultivating 'indy'-led 'revolutions' in a broader momentum for change at the more-than-local level which has whet an appetite to open up conduits for greater levels of democratic representation. These trends may help to create room for political innovations, such as Flatpack Democracy, to germinate.

However, Flatpackery has also been co-opted into other political and environmental change agendas and has been deemed 'complex' to scale up, including by its founding father Peter Macfadyen. Though the movement did not quite have the results that it expected in terms of influencing both political regimes in Frome and beyond, it does not mean the project should be abandoned. Nor does it mean that Frome doesn't offer an aspirational blueprint for others to follow to agitate the democratic parameters of local councils. However, it does suggest that the pull of the groundswell of the wider environmental XR movement and its New Economy affiliates might prove to be a more powerful vehicle with which to align the democratic representation ideals of Flatpack and others to transition towards a wider climate democracy.

The following chapter appraises the objectives underpinning this book and outlines new directions for research on social movements and local government to promote relocalised transitions and the replicability of Frome's experience.

References

Bartocci, L., Grossi, G., Mauro, S. G., & Ebdon, C. (2022). The journey of participatory budgeting: A systematic literature review and future research directions. *International Review of Administrative Sciences*. https://doi.org/10.1177/00208523221078938

BBC. (2022). Somerset election result. https://www.bbc.co.uk/news/election/2022/england/councils/E07000250

Burnett, A., & Nunes, R. (2021). Flatpack democracy: Power and politics at the boundaries of transition. *Environmental Policy and Governance, 1–14*. https://doi.org/10.1002/22t.1931

Comer-Schwartz, K., Craig, B., Dixon, T., Gould, G., Jeffery, D., Mason, P., & Williams, K. (2023). A labour vision for community power: Participation, prevention and devolution. https://www.newlocal.org.uk/wp-content/uploads/2023/09/A-Labour-Vision-for-Community-Power-compressed.pdf

Connors, E. C. (2020). The social dimension of political values. *Political Behavior, 42*(3), 961–982. https://doi.org/10.1007/s11109-019-09530-3

Dennison, J. (2015, September). The Other Insurgency? *The Greens and the Election, Parliamentary Affairs, 68*(Suppl. 1), 188–205. https://doi.org/10.1093/pa/gsv035

East Mendip Green Party. (2018). *Green councillors present their vision of a 'Connected Frome'*. https://eastmendip.greenparty.org.uk/news/2018/02/27/green-councillors-present-their-vision-of-a-connected-frome/

East Mendip Green Party. (2022). *Prospect of a new Frome parliamentary constituency*. https://eastmendip.greenparty.org.uk/news/2022/11/18/prospect-of-new-frome-parliamentary-constituency/

East Mendip Green Party. (2023). *Green Party prospective candidate welcomes new Frome parliamentary constituency*. https://eastmendip.greenparty.org.uk/news/2023/02/13/green-party-prospective-candidate-welcomes-new-frome-parliamentary-constituency/

FTC. (2019). *Themes of Resilience projects influenced or instigated by FTC*. https://www.frometowncouncil.gov.uk/your-community/

Green World. (2021). *Interactive map: 'Record results' for Greens in 2021 local elections*. https://greenworld.org.uk/article/interactive-map-record-results-greens-2021-local-elections#:~:text=The%20Greens%20have%20won%20a%20total%20of%20158,from%20350%20seats%20on%20123%20councils%20in%202019

Green World. (2023). Somerset and Frome: Reflections from the recent by-elections. *Green World*. https://greenworld.org.uk/article/somerset-and-frome-reflections-recent-elections

Heine, S. (2012). Social change in progressive political thought: analysis and propositions. *Journal of Political Ideologies, 17*(3), 301–321. https://doi.org/10.1080/13569317.2012.716629

Hulme, M. (2019). Climate emergency politics is dangerous. *Issues in Science and Technology, 36*(1), 23–25. https://www.jstor.org/stable/26949073

IfF. (2018). *Breaking the Mould*. https://madedifferently.wordpress.com/

IFG. (2023). *Local government section 114 (bankruptcy) notices*. https://www.instituteforgovernment.org.uk/explainer/local-authority-section-114-notices

Lee, A. (2020). Greens offer a different view. *LevellerLive*. https://leveller.live/2020/01/09/greens-offer-a-different-view/

Lee, A. (2022). Green candidate welcomes Frome constituency. *The Leveller*. Retrieved November 12, 2022, from https://leveller.live/2022/11/12/green-candidate-likes-boundary-change/

LGA. (2013). The general power of competence: Empowering councils to make a difference. Retrieved from London: https://www.local.gov.uk/sites/default/files/documents/general-power-competence%2D%2D0ac.pdf

Lodder, M., Allaert, K., Hölscher, K., Notermans, I., & Frantzeskaki, N. (2020). Reflexive monitoring guidebook. In *Reflexive monitoring: Using continuous evaluation techniques to adapt your nature-based solution planning process in real-time*. Drift and Connecting Nature Horizon 2020 project.

Macfadyen, P. (2014). *Flatpack democracy: A DIY guide to creating independent politics*. Eco-Logic Books.

Macfadyen, P. (2019a). *The leaves are fluttering everywhere now ... Flatpack-democracy-inspired groups make advances in UK local elections*. https://www.thealternative.org.uk/dailyalternative/2019/5/5/leaves-fluttering-everywhere-flatpack-elections

Macfadyen, P. (2019b). *Flatpack Democracy 2.0: Power tools for reclaiming local politics*. Eco-logic Books.

Macfadyen, P. (2023, Summer). Flatpack democracy—Was it either? *Stir to Action*, (42) https://www.stirtoaction.com/magazine-issues/issue-42

Manes-Rossi, F., Brusca, I., Orelli, R. L., Lorson, P. C., & Haustein, E. (2023). Features and drivers of citizen participation: Insights from participatory budgeting in three European cities. *Public Management Review, 25*(2), 201–223. https://doi.org/10.1080/14719037.2021.1963821

NALC. (2018). *All about local councils*. National Association of Local Councils. Retrieved from London: https://www.nalc.gov.uk/library/publications/800-all-about-local-councils/file

Rankl, F. (2023). *Planning for onshore wind*. House of Commons Research Library. Retrieved from London: https://researchbriefings.files.parliament.uk/documents/SN04370/SN04370.pdf

Ruiz-Campillo, X., Castán Broto, V., & Westman, L. (2021). Motivations and intended outcomes in local governments' declarations of climate emergency. *Politics and Governance, 9*(2), 17–28.

Salvage, L. (2023, December 15). How extinction Rebellion is democratising to decarbonise. *Friday*. https://www.compassonline.org.uk/how-extinction-rebellion-is-democratising-to-decarbonise/

Somerset Council. (2023). *A new council for Somerset*. https://www.somerset.gov.uk/council-and-democracy/a-new-council-for-somerset/

Somerset Labour Party. (2018). *Frome by election: 2nd place labour now the alternative voice*. https://somersetlabour.co.uk/news/somerset/frome-by-election-2nd-place-labour-now-the-alternative-voice/

Walker, P. (2023). *Tactical voting in byelections spells bad news for Tories.* Retrieved from https://www.theguardian.com/politics/2023/jul/21/tactical-voting-in-byelections-spells-bad-news-for-tories

Wright, C. (2023). *England local elections 2023: The rise of the independents. England Local Elections 2023.* https://www.bbc.co.uk/news/uk-england-65437109

XR. (2023). XR Democracy Hub. Extinction Rebellion. https://www.xrdemocracy.uk/

CHAPTER 12

Conclusion: Towards Regenerative Development in an Era of Uncertainty?

How we can shape truly 'sustainable' development is a key question for our time. Yet the challenge of how to live sustainably is imbued within a myriad of pathways and competing alternatives. At a practical level, some actors favour tweaks to the existing market economy, whilst others seek more radical, or transformative, solutions. Now more than ever we require a robust framework to guide and critique the type of world we are 'transitioning' towards. Brexit and the desire to 'take back control', the changing landscape of environmental movement, the move towards unitary local governance, rapid technological change, the COVID-19 pandemic, the cost-of-living crisis and climate change are all contemporary drivers on the conditions underlying placemaking and innovative political experimentation.

As well as what we've seen in Frome, these issues are relevant for a global landscape on the need to create, capture and circulate sustainable values that underpin the efforts of us all as we seek to find workable solutions to the issues and opportunities we collectively face. We must be able to navigate these interacting and complex drivers to create the mechanisms to forge more inclusive and innovative placemaking and development models that challenge business-as-usual practices and thinking.

This book set out to investigate the scope for implementing New Economy transitions through mainstream institutions or initiatives at the local scale through a multi-level account of transition dynamics. In doing

A. Burnett, *The Politics of Transition*, Palgrave Studies in Sub-National Governance,
https://doi.org/10.1007/978-3-031-40769-7_12

so, this research responds to Carroli's call to explore the fuzziness between boundary extensions and their configuration at the niche:regime interface (Carroli, 2018). This research was based on an identified need to further explore what happens when *Transition* engages with institutional actors or willingly co-opts itself within the systems they seek to change; for instance, whether 'niche' protagonists possess and retain such an identity if they occupy formal spaces of power and what role the 'regime' has on cultivating and contesting development alternatives. Frome offers an opportunity to explore how the *organisational identity* of *Transition* is carried into spaces of government through the actors involved.

I hope to have provided an in-depth account of placemaking networks on local politics in the innovative case study of Frome and done justice to a journey into the relationship between placemaking networks, politics and planning in enabling (or frustrating) more radical (subversive) forms of sustainable placemaking. By following the relationships between Frome's relocalised placemaking entities, including Sustainable Frome and its related spin-off groups, and how they relate to mainstream development structures, notably the Neighbourhood Development Planning (NDP) process and the town council, we can begin to reveal how two 'tribes' of the New Economy—(i) *Transition* and relocalisation and (ii) government and politics—intersect as potential transition arenas for innovative development alternatives. Below, we take a step back from what the goings on in Frome towards a more bird's-eye view of what experiences there reveal about the conditions of innovative community-led planning and placemaking in other contexts, whether green policies and initiatives might be more readily accepted under the guise of independent politics, and how a move towards regenerative development could reinforce and resource sustainability innovations at the local level.

Flatpack Democracy and Its Potential for Scale Shift Towards New Forms of Political Identities

In Frome, the foundations of an independent political movement were laid in part by the *Transition* movement's own rules to seek to formally engage the local council, which became a wider movement strategy (i.e., counter movement trend; see Avelino, 2017) to influence political actors. At the same time, the *Transition* movement's broader (early) rejection of party politics and formal institutions led, indirectly, to IfF's break into political representation. The philosophical ties between IfF and Sustainable

Frome were carried into the spaces of government through the actors involved; actor identities and practices blended with new formal and informal arrangements (which I refer to elsewhere as *informal institutional capital*; see Burnett & Nunes, 2021).

This also indicates how formalised approaches within alternative, niche networks were directed towards formal institutions and, at the same time, hyper-informal practices softened the governance arrangements within the state. With freedom from party political manifesto constraints, IfF could use *local* artefacts and a promise of change to frame their approach and shift (local) political identities and administrative arrangements—with a large degree of political craft and opportunism. Familiarity strengthened ties between placemaking groups and actors to navigate the state (the town council) when IfF occupied power, reflecting Will's insight that pre-existing ties facilitate connections (Wills, 2016).

There was thus a fusion between government-directed transformative powers and social movement transformative powers (in this case the New Economy) where counter-cultural ideas were ignited through regime-directed ideas and structures. Party political identities were key containers for role boundaries and role conflict, or antagonism.[1] Various perceptions of actor types and organisational culture reveal how othering and assimilation of politicised group identities affected transition dynamics, partnerships and the foundations of reciprocal relationships amongst different actors within 'transition arenas', such as the NDP or the non-party-political group the Independents for Frome (IfF). Diversity—and the finding of a common framework that cut across political identities—was a key strategy to mobilise different approaches to reconfigure local action on civic issues, the pursuit of greater control over local issues and to engender a more creative approach to placemaking. However, the extent of this institutionalisation suggests this is tied to the personality of a given role occupier, thus emphasising how organisational and personal identities filter one another; an area that merits further research.

There were instances *of a politics of niche and regime framing* which indicates a variation—or relativity—to whom or what niche and regime concepts mean. For instance, definitions of the niche and regime were used to frame planning as 'niche' for wider interests and it was also not only the 'regime' that led to constraints in role enactment. By taking a closer look at the intersection and power dynamics between local entities,

[1] Which was present in both the Community Plans, within FTC and the NDP.

the research found that the 'niche' and 'regime' do not necessarily have rigid boundaries but reflect an *intersection of transition regimes* with both emergent and strategic foundations (see Burnett & Nunes, 2021).

So, Is Localism Delivering on Genuine Alternative Visions for Sustainable Placemaking?

We explored whether Frome's Neighbourhood Plan was a potential mechanism (transition arena) to promote relocalised sustainability transitions to penetrate planning and its related socio-technical system, the development sector. Notably, do the rights bestowed to local authorities under the Localism policy agenda offer a mechanism to lock-in relocalised identities within mainstream development processes?

As we saw, how a policy is rolled out sets the tone for alignment of Local Planning Authority (LPA) cooperation. While the planning regime was framed as working within different *phases of transition (i.e., episodes of institutional and ideational orderings)*, a perceived lack of political maturity in the rollout of the new NDP regime indicates a discord with its resultant procedures. The conflation of political and vested interests (i.e., developers) in planning policy was seen to undermine an alignment for a more progressive localism that empowered local actors (see Flyvbjerg, 1998; Cowell, 2013) and questions the role of NDPs and the planning regime as a mechanism for transition. Meanwhile, the disparity of being able to control and "change the world" with a relatively "dull" process indicates the need for planning practice to bridge this gap to mitigate disaffection and to heighten the ability for planning to act as a tool for transition.

Frome's NDP process was challenged on the highly procedural form of 'softer' placemaking objectives and the 'mangle' that 're-scripted' and 'squeezed' intent (see also Parker et al., 2017 and Parker & Street, 2015) as they confronted the 'stripped back' 'realities' of planning regime—thus indicating the *reinforcive* powers of the planning regime (see Avelino, 2017).[2] The process did not lead to any immediate changes to LPA–FTC relations nor influence the wider development sector that hampered relocalised placemaking. However, the rights under localism—or their promise at least—helped to galvanise active and imaginative placemaking actors in the town to pursue genuine alternatives to mainstream development

[2] For instance, by encouraging too small-scale action on low-carbon issues (a similar charge made to *Transitions*; Chatterton & Cutler, 2008).

processes, set within a co-constructed understanding of the DNA of Frome, or what was understood to be its essential identity. This then harboured New Economy solutions to stagnant development issues.

However, the political tensions between the protagonist FTC and district council (Mendip District Council, MDC) were perceived as playing "a different ball game" which caused structural holes, broken links and "brick walls" between niche (relocalised) entities[3] and the LPA (i.e., formal knowledge, systems and culture); this hampered the synchronic potential for alignment on issues, compounded by party-political antagonisms. In some ways new opportunities have since been created as different role holders mediated these ties and changes in land ownership and unitary governance in local government for the county of Somerset have seen the emergence of a radical solution to the perennial local issue of 'what to do with Saxonvale' through the community-led Mayday Saxonvale scheme – a battle reminiscent of the David and Goliath where in a rare case in planning terms, the niche entrant might seem to be winning given the decision of a judge of a judicial review case to quash an scheme supported by the former District Council in support of Mayday campaigners. Though this battle is far from over as the rival developer, backed by Somerset Council, continues to advance their proposals despite the High Court decision. Who will end up as the developer of the site as the unitary council seeks to reclaim its dire financial situation is expected to reach its conclusion in the Summer of 2024 – relocalised, community-led (Mayday) or 'mainstream' developer with no 'connection' to the town (Acorn)?

Sustainable Frome - as a transition initiative—was highlighted as a vital organisation that corralled relocalised ideas, actors, interests and friendships in a safe space that also encouraged role transitions—to varying degrees—for those who engaged with it from more 'mainstream' arenas. Pre-existing ties and reciprocity helped to express and advance relocalised place-based artefacts using the structures of the state to weave in institutionalised forms of alternative governance to support local placemaking initiatives (and beyond). While placemaking through Sustainable Frome members and FTC (often the same people) cultivated ***an ecology of localised potential development alternatives*** (tied to the wider New Economy movement), the NDP did not provide an 'engaged interdependency' (Zhang & de Roo, 2016, p. 2) of these niches:regime linkages; nor did the

[3] The NDP process actually broke some ties between Sustainable Frome and FTC either by turning off a "connection"—through different people-links—or being a dry process presented as divorced as a meaningful vehicle for action.

process engender a reformist agenda for more radical (just) planning (i.e., Miraftab, 2009; Beard, 2003), partly due to the failure of the process to enrol 'ordinary' citizens and transcend perceived gender or class-based divisions. Nevertheless, by actively using the new community rights to embed prefigurative practices, Transition's engagement in planning may well shape the increased generation of such practices by more mainstream actors and inspire other grassroots actors to do the same. Indeed *Transition* itself has morphed and flowed through new environmental activist movements, such as Extinction Rebellion (XR) and now the Humanity Project—a loose federation of New Economy and politically-charged movements seeking genuine democratic participation to influence (environmental) politics from the grassroots up and out.

The identity of independence was a strong meta-identity for several individuals in Frome: for instance, the town's 'own' independent identity, its historical—and future—role as a self-sufficient market town and the relative independence of Sustainable Frome from Transition Network, as well as the personalities of those involved in Frome's environmental politics. IfF was perceived to embody and cultivate such an identity that was not previously reflected by political actors and organisations. Cultural differences were a key factor in setting the parameters of engagement in both placemaking networks and within political and planning regimes. Here, an identity of independence was partially tied to spatial character as well as the affective ties of belonging to a social group; in this case, friendship and identity were carried by both the governance of Sustainable Frome and IfF as actors transitioned into new roles carrying with them shared practices and ideas into the town council or other environmental networks, to varying degrees. FTC's planning instruments[4] sought to root Frome's historic identity as a form of spatial and organisational assimilation, which was then consolidated through resourcing and strategic assets and alterity through various roles within FTC.

For Frome-based actors, engagement on issues was perceived to need to reflect the governance and culture of place. 'Knowing' the town's purpose, and the governance to enable this, was a key platform to create a broader movement for change; this was inextricably tied to a strong sense of individual autonomy by key changemakers in the town. However, despite these insights culture is often not aligned within broader conceptions of the politics of transition which ought to be addressed, given its central role in

[4] That is, the Neighbourhood Plan and the Town Design Statement.

corralling shared personal and organisational identities in Frome. Frome's experience suggests a renewed role of place in crafting sustainability transitions. Place identities were invoked to enrol actors in relocalised placemaking networks, establish protagonist challenge to incumbent power and stabilise strategies that sought to solidify such identities within formalised tools and arenas (i.e., the NDP and Town Design Statement).

However, the mantra of IfF councillors that FTC is "still only a town council" indicates a recurrent lack in formal powers arising from localism. Moreover, some of the structures created within FTC are at the mercy of future councillors to change the administrative arrangements, should IfF lose power.[5] Other (landscape) forces to oust political incumbents was seen by some as the only way to expedite change and a "radical shakeup" within and between local government. With the new administrative arrangements in place under a unitary authority, this highlights an exciting avenue for future research that assesses the story of Frome's relocalised, independent politics over time as there are significant structural and resource re-alignments which may affect how community powers are demanded and utilised (such as community hubs), not just in Frome and Somerset, but nation-wide in the wake of increasing moves towards unitary authorities in an effort to make 'efficiency gains' in the representation of local government structures.

However, a fundamental sticking point to the reconstitution of political and relational ties between tiers of government and community representation is the massive budgetary deficits that face local authorities, with a growing number facing extensive cuts in an effort to balance the budgetary books. Somerset Council had already inherited financial difficulties from its district predecessors and these continue to worsen at the time of publication, with few other places to find funds in the coffers. Localism was criticised by some for expecting communities to step up to support services that ought to be delivered by local authorities. While attempts such as introducing Local Community Networks in Somerset hold promise for greater coordination between grassroots and unitary government, without sufficient resourcing they will unlikely be able to function with the desired effect. Initiatives such as the Frome Neighbourhood Network (FNN) and its attempt to broker a 'neighbourocracy' may well meet the dire need to find creative and workable ways to do things differently. However, brokering these ties requires genuine collaboration and shared understanding of optimal ways of working, beyond divisive political or other identity boundaries.

[5] For instance, revoking any changes by IfF to FTC's standing orders.

Frome's case has revealed important insights on the locally-situated degrees of broader social movement ties—to *Transition* and the New Economy and the non-party political 'indy' movement—within Frome's placemaking networks and how broader trajectories (notably localism) may influence these dynamics. Meanwhile, exploring the duality of the 'niche' and the 'regime' within placemaking networks and both formalised and informal structures provides an extended view on the remit of planning on placemaking to solicit conditions for (relocalised) sustainable development. In particular, this study has illustrated how transition arenas are containers for complex power relations (i.e., actors enrolled or excluded from such processes), the influence of social and affective ties on the governance of these arenas and their intersection, the relationship between procedure, culture, personality alignment and autonomy, and the temporality of tools and strategies to provoke transitions.

Yet the experiences of planning and the NDP showed much greater regime tendencies than the political regime, which IfF were able to reconstruct at a local level. The findings echo much research on the restrictive role of planning and a disconnect between the regime and potential coevolutionary approaches between niche and regime to craft systemic transitions (see Carroli, 2018) or challenge elite capture of planning instruments (see Purcell, 2009). There is clear resonance with extant accounts of the NDP process as containing more radical intent, outlined by planning theorists (e.g., Parker & Street, 2015; Parker et al., 2017). IfF had hoped to use the NDP to establish their own planning rules (i.e., Avelino's concept of radical niche regimes) or forming new radical assemblages (Forester, 1989) to contribute to relocalisation and force developers and the LPA to re-prioritise local interests in development decision-making.

As a mechanism for influence, the NDP was perceived to lack system change powers and proved an intractable route to changing the rules of engagement with the planning regime. However, it was also one of the first 'Frontrunner' Neighbourhood Plans supported by the government to be pioneers in the process, and with new processes there are teething problems. With the changes to the NPPF in December 2023 giving greater powers to neighbourhood plans as they weigh in as a 'material consideration' in planning decisions—underpinned by the Levelling up and Regeneration Act 2023—it may well be that policy is aligning to meet grassroots neighbourhood planners in the middle. Yet the politics of 'giving communities power' is not straightforward, as we have seen. Policies change and people switch their voting preferences. The future is never easy to predict, particularly in a context of a climate and ecological emergency.

Looking at Frome's Neighbourhood Plan in isolation is not sufficient to understand wider transition dynamics. An adversarial approach was invoked by FTC as a strategy to claim the discursive space occupied by volume housebuilders and the planning regime, seeking to infiltrate ideas into the spaces of planning to embody new approaches within the system. The relative degree of flexibility, curtailment and entitlement to participate or influence processes within the NDP process echoes Berget et al.'s conception of weak, medium and high ties structural couplings with differing degrees of resonation or structured practices and arrangements (see Bergek et al., 2015).

Reflecting on the capacity for independent 'Flatpack Democracy' to resolve some of these tensions, a more coalitional politics is becoming more prominent recently which could help to scale out political innovations. For instance, More United have suggested that left-leaning parties work together to put up the candidate most likely winning to prevent the Conservatives from succeeding (i.e., as Frome did in its forging of ties with the Greens and local progressive actors nationally and internationally). However, the champion of Flatpack Democracy (Peter Macfadyen)—to my surprise—effectively cut ties with the movement he nurtured, feeling that cultivating radical change in the contours of the (local government) state was not the best place to effect change. The structures around local government are heavily institutionalised. Power does not always reach the very local parish and town councils. Even when it does people don't always know what to do with it and how to retain a radical edge and avoid slipping into business-as-usual. Even in Frome, IfF has in some ways been a victim of its own democratic success. It seems that now few party-political candidates stand against IfF at town council elections—apart from the seasoned local politicians—challenges Iff and if they do they have to play political chameleon and pretend to be an 'independent'. In the 2022 local election, IfF co-opted its councillors in all but one of Frome's eight wards, so IfF candidates didn't even need to stand and be voted in. Wasn't co-option one of the democratic challenges raised by Macfadyen in the story of why a more radical and representative non-party-political politics was needed?

To respond to this call, I suggest the notion of *placemaking transitions*—the locally situated dynamics of place and politics in networked transition arenas—to capture how the politics of place influences sustainability transition arenas. Such a definition can help to explore the role of placemaking and transitions as mutually reinforcing bodies of research and build an evidence base for what works, or doesn't work, to realise more sustainable communities. Planning transitions theory could be a branch of both planning scholarship and sustainability transitions studies which work simultaneously to fully explore politics, autonomy, culture, identity, place and processes to plan for transitions. For this, I will flesh this idea in future research, but I invite others to work with me in this endeavour to explore these issues in more detail through similar comparative work.

Reflections on Future Research and Limitations

It is important to note that this study focused explicitly on the influence of *Transition* on relocalised placemaking networks and their influence on the structures of the state; it did not interview those outside of this network (other than the LPA as an instance of possible 'regime' resistance), such as those who may be classified by some as "Old Frome", to substantiate claims of the effects of possible gentrification and political identities. Nor did it interview developers to explore the extent to which these entities supported or resisted relocalised placemaking. This study—though attempting to engage them—did not corroborate the findings with party political actors. Though recent conversations did seek to understand the perspectives of Green Party representatives and those championing radical solutions to community-led development in the town. These limitations would provide fruitful future research to substantiate the findings of this particular project and to explore whether the proposed intersecting regimes of transition is relevant to these actors in similar ways to those interviewed in Frome, or not.

Whilst there is explicit endorsement of many voters in Frome of IfF, the extent to which people have transcended their previous political identities and what conditions might activate their previous political identities is required to understand IfF's role in placemaking transitions. It would be revealing to explore which values members of Transition/Sustainable Frome/ XR / Humanity Project actors would compromise on, which would indicate the commonalities that bind the movement and how 'external' factors to the movement have responded to the capacity to influence key moments of change. In particular, how the Advocacy Coalition Framework (Sabatier, 1998) can be unpacked to explore key 'identity switching factors'. Multiple case comparisons would also allow identify the reach of similar causal properties across different cultural and institutional environments and whether the empirical regularity remains similar or results in a typology of approaches and lead to a more robust account of change processes across different scales (Tsang, 2013, p. 6).

Such an approach might explore whether there are commonalities in the evolution of organisational emergence in independent political groupings, or under the New Economy umbrella. In particular, what organisational traits are carried or mutated across these social movements? From where do they draw their identity markers and which resources do they have to become an identity regime? A comparative approach between places that share similar identity frames (such as Bristol, Brixton, Brighton, Frome and Totnes) could explore the role of *Transition* in influencing placemaking dynamics and the similarity and differences in how identity networks are formed and eco-identities consolidated.

Indeed, planning was seen as a disruptive force against potential gentrification patterns; the outcomes of which would necessitate further research with inhabitants of social (affordable) housing and their engagement with the town's placemaking networks. This would require a more in-depth study on the voting preferences of IfF voters for each ward and the reasons why they voted for these candidates. Moreover, the socio-economic profile of membership to other local groups would also reveal if there were specific class divisions and if these were being accelerated or reconfigured through IfF's politics. Moreover, the representation of class and gender amidst growing independent political councillors would illuminate whether there were gaps or bridges in class, gender and representation in new political spaces. Who and what is causing the 'Green Wave' in Frome and how do these new voting patterns cut across the intersectionality of other social and economic identities? Is it the "New Frome" and "Down from London"ers that are causing this political sea change, or is something more fundamental shifting within people's appetite for change?

The findings highlight a need for greater attention to the cultural qualities of placemaking as a basis for social ties that support inclusive and emergent governance. However, Frome's cultural identity was reproduced differently as economic drivers influenced the socio-economic profile of those moving to the town, raising tensions about class and gentrification in transitions processes and what might be the switching factors that determine Frome's DNA in its socially constructed genetic makeup (and what is and isn't included). The dynamics and tension between old and new instigated role change and organisational evolution (not just incumbent and protagonist), are reflected in a the parent–child relationship between

Sustainable Frome and its associated spin-off groups. Other examples include perceived disparities in the mindset of the younger and older generations, and the draw of family influences, indicating the need for a developmental approach to understanding patterns of transition, emergence and affective ties.

Concluding Remarks

There have been at times signals of a growing rejection of party politics which might indicate fissures in the institutional arrangements of party politics and a possible de-alignment of political identities. However, as the British public go to the polls in July 2024 party politics again seems to be pervasive in shaping national priorities. The 2019 May local elections saw an unprecedented rise in the number of independent political councillors elected at town and district level.[6] The conditions for their rise and potential success and challenges in influencing government and other entities is another vital area of research. How the 'indy' movement evolves to influence the national level I thought was one of the most fruitful areas of research on British politics today to cultivate an authentic coming together of people committed to making where they live better, whatever their historical or current political pursuasions. However, the experiences of Macfadyen in abandoning this particular movement and instead shifting his efforts to a more collective effort to forge radical democratic transition arenas outside of formal political systems is telling of the extent to which local governments can be the transformative vehicle promised by some of Macfadyen's early works on Flatpack Democracy. How the Humanity Project fares, with some of the radical social movement heavyweights of our time working together for shared democratic ends, will be one of the most fascinating political agendas of the 2020s. In particular, how it seeks to bridge ties with government rather than seek to subvert it all together.

Understanding Frome's placemaking networks and efforts to foster change within the system has enabled planning and the NDP process to be considered within a broader package of relocalisation and transition. Frome's experiences clearly demonstrate both strategic (steer) and emergent qualities that interface with wider phenomena. Narratives of

[6] See https://www.thealternative.org.uk/dailyalternative/2019/5/5/leaves-fluttering-everywhere-flatpack-elections?rq=leaves for instances of 'indy' gains in the May 2019 English local elections and https://www.somersetlive.co.uk/news/somerset-news/local-elections-results-2019-mendip-2831123 for the Mendip district results (both last accessed 10.6.19).

relocalisation and independent politics are clearly influencing Frome's placemaking and its reputation as it becomes increasingly renowned for its thriving independent identity, politically and economically which has helped to orchestrate multiple identities and their changing alignments to echo a broader movement for a New Economy. Relocalisation reached a degree of maturity and power in new administrative arrangements (i.e., branching from community-based organisations into the state), fuelled by the 2011 Localism Act, to enable the conditions for successful place-making (and governance arrangements) to take off in a subversive way. More recent events suggest that due to this groundwork in other aspects of Frome's placemaking, the LPA has begun to take a more pro-active approach towards local interests and development priorities. Relocalised energies did not dissipate and circulated at a local level (primarily within FTC and latterly XR) to create an energy and momentum for change. With unitary governance and a change of tactic towards more subversive democratic outcomes at a local level, this combination could prove an interesting new political landscape for sustainability transitions to take form. However, Somerset Council's decision to continue back the Acorn Saxonvale scheme or change tack and back the alternative Mayday scheme as it battles its own financial woes will be telling in the reach of locally-led development alternatives amid 'business-as-usual' development.

Measurement and attribution of sustainable impacts are accelerated through digital architecture and data within the *GovTech* agenda to deliver new and efficient ways to engage communities in service delivery; a drive towards digitalisation could help to connect data and services, as well as to invite Small and Medium-Sized Enterprises (SMEs) and other entrepreneurs into the delivery of hybrid public–private services.[7] Such technologies offer a significant contribution to the DigitalGov sector by opening up public bodies to cutting-edge digital architecture and reconfiguring management and public services, structures and associated framework for evolving capacities in the workplace and the value chain, as well as the political communication structure within and between government (Filgueiras & Silva, 2021). This also opens up the space for new actors to enter in that may not had a role in shaping local government services before, raising new questions for accountability and citizen–state relationships. A clear roadmap is required that considers the resources and political

[7] This is encouraged by government by events and initiatives such as the GovTech summit and forums such as TechUK: https://www.techuk.org/developing-markets/central-government.html

buy-in of implementation, which may conflict with existing administrative regimes. It also suggests that the impact of socio-technical innovation on people's capacity for engagement should be explored in more depth, including its effect on breaking open planning and politics or fragmenting energies and momentum for change as people "blank out" competing priorities in an ever-distracted (online) world. People are also seeking a more mindful approach to how they relate to the world. How digital technologies might play a role in a more mindful politics remains to be seen.

More recently, there have been increasing discursive moves away from resilience towards the concept of regenerative economies; that is, encouraging the positive reinforcement of multiple social, environmental and economic benefits. For instance, Platform Design Toolkit is exploring how to support Regenerative Platforms (PDT, 2019). Sustainability platforms and emergent supply chains could be further enhanced through greening incentives. For instance, where actors are motivated to choose greener choices by rewarding those who engage with more sustainable choices that can promote a circular economy through an offer of green rewards (Zarra et al., 2019). These have promising links to forging beyond Gross Domestic Product (GDP) and economic-centred models (a long-claimed challenge of more 'progressive' commentators) towards supporting alternative forms of value and shared governance, such as those encouraged within Transition Town circles and the New, or Next, Economy.

Examples of these values in action can be found in platforms that aim to localise supply chains, citizen-focused digital tools that bring citizens into the fold of public service delivery and development priorities. For instance, Loconomy has a "focus on local self-determination and self-reliance, including self-determined distribution, in-person community building supported by on-platform groups and fora to share relevant information for the local community".[8] Various platforms have emerged that enable more interactive planning engagement. For instance, Commonplace suggests that its platform "[g]ive[s] your community an easy, accessible way to get involved in the planning process".[9] In this particular tool, users see development applications through a social media post which invites engagement and assesses user credentials to "reach the 'silent majority'".

In addition, were there to be greater (financial or non-monetary) incentives to encourage the collection, disclosure and sharing of data, these

[8] https://www.rizomafoundation.org/loconomy

[9] https://www.commonplace.is/

would necessarily need to promote trust and engagement in data governance, all of which are likely to differ across regions. This is something I have written about - where digital platforms and new nature-positive mechanisms to cultivate environmental and social value can support the measurement and exchange of regenerative value (Burnett, 2022a; Burnett 2022b; Bream McIntosh et al., 2023). There are also many competing objectives and priorities that need to be measured and considered; as Filgueiras and Silva (2021) point out, data governance itself can lead to political conflict as new relationships are formed with citizens and mediated through often third-party technology companies. Smart city and other digital projects can sometimes lead to policy failure as well as innovation. As such, data models should account for how values surfaced and were contained or co-evolved through governance processes that stimulate or block sustainability pathways (see Bocken et al., 2013).

In addition, scale and networks should consider how different rural–urban configurations affect innovation in digital and/or innovative supply chains; evidence suggests that smaller cities and municipalities have more tight-knit relationships that may affect how quickly innovations can take root, particularly where these leverage insights across multiple government departments or embrace community-driven problem solving (Mansour, 2021).

Seeing data as a form of digital commons is one way of conceptualising the role of data and the citizen-local government relationship, which could generate new relationships that foster more inclusive and collaborative approaches (Filgueiras & Silva, 2021). How democracy, agency and authenticity unfold in an age of rising artificial intelligence remains to be seen. However, digital innovations such as blockchain, and their role in potentially documenting the value of 'sustainability' as it is exchanged across different democratic and enterprise systems is a fascinating area of future research, and one that may well contribute towards more regenerative and inclusive democracies - but that too remains to be seen.

References

Avelino, F. (2017). Power in sustainability transitions: Analysing power and (dis) empowerment in transformative change towards sustainability. *Environmental Policy and Governance, 27*(6), 505–520. https://doi.org/10.1002/eet.1777

Beard, V. A. (2003). Learning radical planning: The power of collective action. *Planning Theory, 2*(1), 13–35. https://doi.org/10.1177/1473095203002001004

Bergek, A., Hekkert, M., Jacobsson, S., Markard, J., Sandén, B., & Truffer, B. (2015). Technological innovation systems in contexts: Conceptualizing contextual structures and interaction dynamics. *Environmental Innovation and Societal Transitions, 16*, 51–64. https://doi.org/10.1016/j.eist.2015.07.003

Bocken, N., Short, S., Rana, P., & Evans, S. (2013). A value mapping tool for sustainable business modelling. *Corporate Governance: The International Journal of Business in Society, 13.* https://doi.org/10.1108/cg-06-2013-0078

Bream McIntosh, O., Burnett, A., Feldman, I., Lamphere, J., Reuter, T., & Vital, E. (2023). The Role of Sustainability Knowledge-Action Platforms in Advancing Multi-Stakeholder Engagement on Sustainability. *Data & Policy, 5*(E33). https://doi.org/10.1017/dap.2023.27

Burnett, A. (2022a). Operationalising performance measurement and incentives for 'regenerative value' to improve net sustainability outputs. In G. Vélez-Cuartas & O. Romero-Goyeneche (Eds.), *Transformative Metrics: Contributions to the Studies for Monitoring and Evaluating How Science, Technology, and Innovation Can Address Social and Environmental Challenges.* Medellin, Colombia: University of Antioquia.

Burnett, A. (2022b). Securing a regenerative and just planning system. *Town & Country Planning, 91*(2), 123–128.

Burnett, A., & Nunes, R. (2021). How well did the Flatpack Democracy 2021 campaign do? This report shows it's still a better future for citizens than traditional politics. https://www.thealternative.org.uk/dailyalternative/2021/11/15/flatpack-democracy-2021-how-did-they-do

Carroli, L. (2018). Planning roles in infrastructure system transitions: A review of research bridging socio-technical transitions and planning. *Environmental Innovation and Societal Transitions, 29*, 81–89.

Chatterton, P., & Cutler, A. (2008). *The rocky road to a real transition: The Transition Towns movement and what it means for social change.* http://trapese.clearerchannel.org/resources/rocky-road-a5-web.pdf

Cowell, R. (2013). The greenest government ever? Planning and sustainability in England after the May 2010 elections. *Planning Practice & Research, 28*(1), 27–44.

Filgueiras, F., & Silva, B. (2021). *Assessing data policy by institutional analysis development framework.* Paper presented at the conference proceedings for a Panel on Policy Measures and Institutional Arrangements to Govern Data-Driven Innovation for Sustainability, the 5th international conference on Public Policy, Barcelona, July 6–8.

Flyvbjerg, B. (1998). *Rationality and power: Democracy in practice.* University of Chicago Press.

Forester, J. (1989). *Planning in the face of power.* University of California Press.

Mansour, K. (2021). How to foster innovation hubs outside capital cities. https://earlymetrics.com/how-to-foster-innovation-hubs-outside-capital-cities/

Miraftab, F. (2009). Insurgent planning: Situating radical planning in the global south. *Planning Theory, 8*(1), 32–50. https://doi.org/10.1177/1473095208099297

Parker, G., Lynn, T., & Wargent, M. (2017). Contestation and conservatism in neighbourhood planning in England: Reconciling agonism and collaboration? *Planning Theory and Practice, 18*(3), 446–465.

Parker, G., & Street, E. (2015). Planning at the neighbourhood scale: Localism, dialogic politics, and the modulation of community action. *Environment and Planning C: Government and Policy, 33*, 794–810. https://doi.org/10.1068/c1363

PDT. (2019). 'The platform design toolkit' Evolving Service design & business modeling in the age of ecosystems. https://platformdesigntoolkit.com/toolkit

Purcell, M. (2009). Resisting neoliberalization: Communicative planning or counter-hegemonic movements? *Planning Theory, 8*(2), 140–165. https://doi.org/10.1177/1473095209102232

Sabatier, P. A. (1998). The advocacy coalition framework: Revisions and relevance for Europe. *Journal of European Public Policy, 5*(1), 98–130.

Tsang, E. W. K. (2013). Case study methodology: Causal explanation, contextualization, and theorizing. *Journal of International Management, 19*(2), 195–202.

Wills, J. (2016). Emerging geographies of English localism: The case of neighbourhood planning. *Political Geography, 53*, 43–53. https://doi.org/10.1016/j.polgeo.2016.02.001

Zarra, A., Simonelli, F., Lenaerts, K., Luo, M., Baiocco, S., Ben, S., ... Kilhoffer, Z. (2019). *Sustainability in the Age of Platforms.* Retrieved from Brussels: https://www.ceps.eu/wp-content/uploads/2019/06/Sustainability-in-the-Age-of-Platforms-1.pdf

Zhang, S., & de Roo, G. (2016). Interdependency of self-organisation and planning: Evidence from Nanluoguxiang, Beijing. *Town Planning Review, 87*(3), 253–274.

Appendix A: A Summary of Key Tokens in Frome's Relocalisation Networks

Some key tokens of relocalisation used in the town are captured in Table A1.

Table A1 Visual Identities (c/f Chap 5.)

Image No.	*Visual Identity of*	*Summary*
1	Flatpack Democracy	A handbook of do-it-yourself politics based on Frome's experience written by the co-founder of IfF (Macfadyen, 2014).
2	Independents for Frome	A group of independent (town) councillors who have been in power since they were elected in 2011.
3	Frome Town logo	The Frome town emblem used in a number of community logos.
4	The bearded man	A recent parody of the type of 'trendy' 'New Frome'—incomers—who are attracted to the town.
5	Creative activities—Frome Street Bandits	A group of Frome-based musicians, one of many artistic and musical groups in the town. See http://www.fromestreetbandits.co.uk/

(*continued*)

A. Burnett, *The Politics of Transition*, Palgrave Studies in Sub-National Governance,
https://doi.org/10.1007/978-3-031-40769-7

Table A1 (continued)

Image No.	*Visual Identity of*	*Summary*
6	Caption from IfF 2015 election campaign video: 'For the People, by the People'	IfF marketing co-opting the Frome emblem.
7	News article on launch of Frome's ambition to become UK's first fossil fuel free town	Intention launched by FTC for the town to become the fossil fuel free by 2046; subsequently superseded by a commitment to become net-zero by 2030.
8	Re—love—ution	Used on the Flatpack Democracy blog https://www.flatpackdemocracy.co.uk/
9	Frome Community Fridge	Established in 2016 with support from FTC to reduce food waste by encouraging donations from local food retailers. The 'Community Fridge' concept has since become a commonplace feature of many UK communities https://www.communityfridgemap.org.uk/.
10	Frome Independent	Established in 2013, the biggest independent market in South West England. It aims to be 'more than a market' and offers monthly themed markets (exc. winter months). On Frome Independent market days, the high street is closed to traffic and has over 10,000 visitors.[a] It is a core part of the creative cultural capital of the town; it seeks to be 'more than a market' and to enable other cultural and social forms of interaction to arise. For instance, partnership with the private sector, culture, art, as a social experience (to be shared with friends and family) and celebrating independence itself. The market, like Frome's politics is bold and ambitious—for instance, creating Frome-on-Sea for one of the themed Frome Independent market days.
11	Mayoral chains	An array of alternative Mayoral chains commissioned by Peter Macfadyen during his time as Mayor of Frome. These were symbolic of related social and environmental initiatives.[b]
12	The Share Shop	Established in 2015 as a joint initiative between FTC and Sustainable Frome to encourage re-use and mending of household items to promote a circular economy
13	Fair Frome Housing	Established in 2016 to explore alternative ways to support affordable housing, such as community housing and linked apprenticeships for youth unemployed, and house sharing. Members include IfF councillors.

(*continued*)

Table A1 (continued)

Image No.	*Visual Identity of*	*Summary*
14	Sustainable Frome	Frome's Transition Initiative, established in 2006 to support local environmental initiatives; the group formally became part of the Transition Network in 2008.
15	Keep Frome Local	A campaign, with roots in Sustainable Frome, to prevent a supermarket being built at the redundant Saxonvale site. Rob Hopkins, co-founder of Transition Network, spoke at one of its principal events, 'Independence Day' in 2012, held at the Wesley Methodist Church.[c]
16	Keep Frome Clean	An FTC initiative to encourage multi-stakeholder engagement so that Frome is a clean place to live and work.
17	Caption from IfF 2015 election campaign video: Whatcombe fields	An initiative launched in 2015 by Save Open Spaces (SoS) Frome—a consortium of representatives of local environmental groups—as a successful crowdfunding campaign that enabled the community purchase of a local field to save it from development. A total of £300,00 was raised, which included a £35,000 contribution from FTC.
18	Caption from IfF 2015 election campaign video: co-wheels outside the Cheese and Grain	An initiative by FTC to bring electric car sharing to the town. Now also served by E-car club.
19	Catherine Hill, Frome	A quintessential Somerset street in a market town, now host to an array of independent retail which was rehabilitated since it had been run down in the 1990s and early 2000s.
20	Edventure	'A school for Community Enterprise' supporting youth and other community initiatives.[d]
21	Saxonvale	A mostly derelict 4.76-hectare site in a central location in the town. A source of contention over whether a supermarket should be built on the site (debated by 'Keep Frome Local' who were opposed to a supermarket on the site and 'Frome for All' who supported a supermarket). More recently, Saxonvale was the arena where development alternatives played out again under the 'Mayday Saxonvale' campaign, which sought more socially inclusive, creative and environmental design as opposed to the scheme put forward by an alternative development proposal. See https://maydaysaxonvale.co.uk/

(*continued*)

Table A1 (continued)

Image No.	*Visual Identity of*	*Summary*
22	One Planet Living (OPL)	An initiative developed by Bioregional and WWF to set out the conditions necessary to live within the planetary boundaries of one world.[e]
23	In the Middle of Nowhere, Centre of Everything (logo)	A phrase used by the Frome Independent to indicate the high levels of activity in the town in its rural setting.
24	Frome Cohousing	Established in 2009 to seek sites for a cohousing community.
25	The Cheese and Grain	A restored building in central Frome that is host to a number of live music and community events.[f]
26	Forward Space/The Old Church School	A co-working, hotdesking facility for 'Independent and Creative Thinkers' in the centre of Frome—and latterly expanded to nearby towns and cities and now includes collaboration between Forward Space and Mendip District Council (MDC) to explore flexible work units across Frome.
27	Image of the front cover of Frome's NDP	Frome's Neighbourhood Development Plan (NDP) was initiated in 2012 as a response to a call for Frontrunners under the Government's localism agenda.

[a]http://www.thefromeindependent.org.uk/about/ (last accessed 25.9.17)

[b]http://www.frometowncouncil.gov.uk/the-mayors-chain-exhibition/ (last accessed 25.9.17)

[c]http://www.frometimes.co.uk/2012/09/12/keep-frome-local-launches-independence-day/ (last accessed 25.9.17)

[d]https://edventurefrome.org/ (last accessed 11.1.18)

[e]http://www.bioregional.co.uk/oneplanetliving/ (last accessed 25.9.17)

[f]http://www.cheeseandgrain.com/ (last accessed 11.1.18). More recently attracting well-known international bands such as the Foo Fighters

Appendix B: Sustainable Frome's 'Objects'

Sustainable Frome Objectives

Sustainable Frome CIC will work with individuals, groups, organisations and businesses, primarily in Frome and surrounding areas, to enable current and future citizens to live lives that are pleasurable, healthy, purposeful and just, without overspending on the environmental budget. The organisation aims to increase the overall resilience of the people of Frome to future environmental, economic and social shocks and challenges.

Specific objects are to support, develop, initiate and manage community projects and interactions that will

1. regenerate biodiverse, resilient and vibrant local economies, food systems and land use;
2. seek local and sustainable economic and energy solutions that enable local people to commit to low carbon lifestyles and energy efficiency;
3. lead to transport initiatives and opportunities that bring practical solutions to reduce environmental pollution, improve public health and counteract the impact of increasing fuel costs;
4. reduce current levels of finite resource use, and minimise the production of waste;

A. Burnett, *The Politics of Transition*, Palgrave Studies in Sub-National Governance,
https://doi.org/10.1007/978-3-031-40769-7

5. reduce excessive global trade and consumption to "live simply so that others may simply live";
6. enable alliances to other sustainable communities, networks and movements which recognise the links between sustainability and social justice, and encourage engagement with relevant national and international campaigns and movements;
7. lead to creative and artistic outputs, emphasising the importance of their work in community strengthening, education and campaigning;
8. promote and encourage the acquisition of new skills in activities that enhance sustainability and strengthen community.
9. recognise links between health and sustainable communities, forming practical alliances with complementary, alternative and conventional health practitioners;
10. develop recognition that financial wealth does not directly equate to happiness and ensure maximum opportunities for personal growth and expression to play a central role in sustainable community development; and
11. encourage full engagement with younger people on issues of sustainability.[1]

[1] http://transitionfrome.org.uk/about-us/the-objects/ (last accessed 11.1.19).

Appendix C: IfF's Ways of Working

A. Burnett, *The Politics of Transition*, Palgrave Studies in Sub-National Governance,
https://doi.org/10.1007/978-3-031-40769-7

"The noble art of losing face will one day save the human race". Piet Hein

These Values and Guidelines have been drafted by the group of 17 independent individuals elected to Frome Town Council in May 2015, using Piet Hein's mantra as a guide to a values-based approach to town council politics. They are based on the original Ways of Working adopted in 2011.

Five Core Values

Independence. We will each make up our own mind about each decision without reference to a shared dogma or ideology.
Integrity. Decisions will be made in an open and understandable manner. Information will be made available even when we make mistakes, and everyone will have the opportunity to influence decisions.
Positivity. We will look for solutions, involving others in the discussions, not just describe problems.
Creativity. Use new, or borrowed, ideas from within the group and the wider community to refresh what we do and how we do it.
Respect. Understand that everyone has an equal voice and is worth listening to.

We will adhere to these values by challenging ourselves and each other to:

- Avoid identifying ourselves so personally with a particular position that this in itself excludes constructive debate.
- Being prepared to be swayed by the arguments of others and admitting mistakes.
- Be willing and able to participate in rational debate leading to a conclusion.
- Understand the value of constructive debate.
- Accept that you win some, you lose some; it's usually nothing personal and there's really no point in taking defeats to heart.
- Maintain confidentially where requested and agree when it will be expected.
- Share leadership and responsibility and take time to communicate the intention of, and the approach to, the work we undertake.
- Have confidence in, and adhere to, the mechanisms and processes of decision-making that we establish, accepting that the decisions of the majority are paramount.
- Sustain an intention to involve each other and others rather than working in isolation.
- Trust and have confidence and optimism in other people's expertise, knowledge and intentions. Talk to each other not about each other.

http://iffrome.org.uk/our-successes/ways-of-working (last accessed 11.1.19).

Fig. C1 The Independents for Frome Ways of Working IfF (2015)

Appendix D: Made Policies in Frome's Neighbourhood Plan (2016)

A. Burnett, *The Politics of Transition*, Palgrave Studies in Sub-National Governance,
https://doi.org/10.1007/978-3-031-40769-7

No.	*NDP Policy No.*	*Policy Name*	*Policy—Deviations from Previous Version in Italics*
	N/A	Golden Threads	Made plan VCO 1—Vitality VCO 2—Participation VCO 3—Sustainability Submission version VCO 1—vitality VCO 2—sustainability VCO 3—participation
	N/A	OPL	It is recognised that the term sustainability extends well beyond the energy efficiency of buildings. The UK Sustainable Development Strategy Securing the Future sets out five features of sustainable development: living within the earth's environmental limits, ensuring a strong, healthy and just society, achieving a sustainable economy, promoting good governance and using sound science responsibly. In order to provide clarity and policy guidance on future development, the Council is proposing to adopt an overarching sustainable development objective based on One Planet Living as defined by Bioregional and Worldwide Fund for Nature (WWF). One Planet Living (OPL) is a simple concept. We have only one planet and collectively we have a responsibility to live within its capacity. At present, within the UK we are consuming natural resources at a rate that would require the equivalent of three planets if everyone shared our lifestyle. Ten clear principles have been developed to promote happy, healthy and sustainable communities, and these have been translated into a management model:

1	H1	BUILDING A BALANCED COMMUNITY	All new residential development should maintain, provide or contribute to a mix of housing tenures, types and sizes to help support the creation of mixed, balanced and inclusive communities in the Frome Neighbourhood Plan area. Planning applications for residential development should address affordable housing need and housing demand in line with Mendip District Local Plan 2006-2028 Policy DP11: Affordable Housing. ** Reg 14 version had sought affordable housing for applications for 10 or more and contribute to housing diversity, design solutions for strong neighbourhoods and adequate space.
2	H2	BUILDING BY DESIGN	Quality of design is a key consideration for residential development. Proposals which fail to take the opportunities available to enhance the local character and quality of the area and the way it functions or fail to have regard to guidance in the Frome Town Design Statement (October 2015) will be refused.
3	H3	SELF-BUILD AND COMMUNITY HOUSING	As an exception to normal policy for the provision of housing set out in Core Policies 1 and 2 of the Mendip District Local Plan, Community Housing may be permitted adjoining the existing development limit (excluding the area that lies within the parish of Berkley) on sites where development would not otherwise be permitted providing: (a) The development provides an appropriate mix of dwelling types and sizes reflecting identified local need and meets demand based on the current Local Housing Needs Assessment or evidence from local Community Housing Groups and the Town Council. (b) The development will not have a significant adverse impact on the character of the area and local landscape setting. (c) Such proposals should not have an adverse or harmful impact on statutorily protected species or habitats. (d) The land is held in trust as a community asset in perpetuity. Footnote 1 Community Housing is defined as residential development by a group who build on land that is held in common ownership or trust for the benefit of the residents. The following text was removed in the made plan: For community housing outside of the existing development boundary providing there is collective land ownership, diversity of dwelling type and energy efficient. Residents can demonstrate professional and financial capacity; land is held as a community asset; discounted rent at 80% open market value

(*continued*)

(continued)

No.	*NDP Policy No.*	*Policy Name*	*Policy—Deviations from Previous Version in Italics*
4	BE1	PROTECTION OF EMPLOYMENT LAND	In addition to Local Plan Policy DP20, the redevelopment of 'Valuable Employment Sites' for non- employment uses will be resisted. Such development would only be granted in the exceptional circumstances where it can be clearly demonstrated that no demand exists within this area for B1, B2 or B8 uses.
5	TC1	TOWN CENTRE REMODELLING	Remodelling the Town Centre should provide improvements to the public realm. *** Reg 14 policy had sought the allocation of an additional 5ha of employment land. Remodelling of the Town Centre which accords with the following principles will be permitted: • Improve the Town Centre environment for pedestrians. • Reduce the impact of traffic movement from vehicles and re-order the priorities between motorised vehicles, cycles and pedestrians. • Enhance the character and appearance of the Town Centre, taking into account guidance in the Frome Town Design Statement (October 2015) relating to this area. • Provide an improved setting and location for the markets. • Demonstrate that it will cause no deterioration in air quality. *** Reg 14 version had sought to implement option 3 of town centre remodelling feasibility study, subject to significant community involvement.
6	TC2	WESTWAY CENTRE	Redevelopment or expansion of Westway shopping centre will be permitted subject to: • Public realm improvements including access to the Frome River and tree planting. • Sufficient car parking and space for buses being included within the redevelopment to ensure there is no detrimental impact on the vitality of the town centre. • The design takes into account guidance in the Frome Design Statement (October 2015) • There will be no harmful of adverse impact on the biodiversity or protected species.
7	TC3	THE CATTLE MARKET CAR PARK	Landscape and infrastructure improvements to the Cattle Market Car Park that improve this arrival point and enable this area to develop as a cultural destination will be supported. ***Reg 14 version had also sought to explore longer-term management options

8	POS1	THE RIVER CORRIDOR	Otherwise acceptable planning applications that are closely related to the River Corridor environment and approaches to it will be supported where they take advantage of opportunities to improve the River Corridor environment, including access, subject to suitable ecological assessment.
9	T1	INTEGRATED TRANSPORT STRATEGY	The delivery of integrated sustainable transport options is a key aim of the Frome Neighbourhood Development Plan. The priorities for delivering these are: • Completing Frome's Missing Links in the National Cycle Network 24. Any detrimental impact through extension to Frome's 'Missing Links' in the National Cycle Network must be avoided or appropriately and entirely mitigated in order to protect the Mells Valley Ecological Zone of Interest including foraging areas, commuting or migrating flight lines for bats. No development will be permitted where these conditions cannot be met. • Improving access to and along the River Corridor, in line with the Frome River Strategy. • Enhancing the environment around and access to Frome Railway Station. • Providing safe and accessible bus and coach stops • Identifying clear and safe routes for pedestrians and cycling as part of the remodelling of Frome Town Centre.
10	D1	DESIGN IN URBAN LANDSCAPES	Development that fails to take account of the design guidance in the Frome Town Design Statement (October 2015) or the Frome Conservation Area Character Appraisal and Management Proposals (18 June 2008) will be refused.
11	D2	GATEWAY SITE IMPROVEMENTS	Developments that enhances the built environment of the gateway spaces, taking account of guidance in the Frome Town Design Statement (October 2015), will be supported.
12	D3	VISUAL IMPACT ON SKYLINE	Applications for new buildings and highly visible extensions to existing properties in the skyline areas will be required to demonstrate that there will be no unacceptably detrimental impact on the skyline in these skyline areas. ***Examiner report required a visual map of skyline areas.

Appendix E: Deleted Policies in Frome's Neighbourhood Plan (2016)

Various Policy Intentions Were Cut from the Plan

NDP Policy No.	*Policy Name*	*Policy (Summarised)*
H4	Delivering Major Projects: (up to v3)	Developments over 100 houses should set out a management plan for residents' interaction with wider Frome population; including meeting OPL, diverse housing mix and home-working, pedestrian and cycle links, 5% self-build allocation
H5	Energy Efficient Homes (up to v3)	Encouraged to exceed Code 3 Sustainable Homes, consideration of reduction in CIL for energy efficient houses
H7	Future Growth Area:	Frome Town Council will ensure The Mount (as identified as a growth area in the Mendip Local Plan) will be bought in accordance with the NDP.
BE3	Sustainable Development (v1)	Development over 1000 m^2 of employment floor space complies with OPL
TC2	St Catherine's and Badcox: (up to v3)—a non-policy objective in the made plan	No more than 4% of retail change of use in these areas.

(*continued*)

A. Burnett, *The Politics of Transition*, Palgrave Studies in Sub-National Governance,
https://doi.org/10.1007/978-3-031-40769-7

(continued)

NDP Policy No.	*Policy Name*	*Policy (Summarised)*
TC5	Saxonvale (up to v3)—a non-policy objective in the made plan	Redevelopment of site based on 2005 Planning Brief; work with Mendip District Council, landowners and community and update Brief to deliver: exemplar low-carbon living, 300+ energy efficient, diverse mix of homes, incubator space for SMEs, retail, hotel and leisure facilities, training facilities for Somerset Skills and Learning, artisan workshops and creative space, urban park alongside river, improved traffic management.
POs2	Public open space (up to v3)	Development of public open space will not be permitted unless it promotes access and use of public land and buildings, not cause the loss of such areas, developments over 100 homes/1000 m^2 commercial floor space will detail protection of open space/wildlife around site area, introduction of supplementary planning document on green infrastructure to identify deficiencies in open space and leisure
D4	Tree Protection (up to v3)	Development should not contribute to loss or damage of woodland and have landscape and ecological value; three trees should be planted for each new dwelling or one tree for each car parking space or one tree per 50 m^2 of gross floor space in non-residential either on-site or on another site subject to development.
T2	Travel Plans: (up to v3)	Residential schemes over 10 units, new employment and missed use over 1ha, schools and colleges, health and leisure centres and public buildings will produce a sustainable transport plan

Glossary

Article 4 Directions Article 4 directions are made by the Local Planning Authority (LPA), which removes or restricts the use of otherwise permitted development rights.

Community Planning Plans that communities can create to represent their vision for local development priorities, but these may not become statutory planning documents, unless adopted by the LPA as supplementary planning guidance.

Independent Politics Elected officials or candidates who reject party politics and seek more freedom to represent of their personal or community views.

Local Planning Framework A set of policies, guidelines and regulations that govern land use and development within a specific local area.

Neighbourhood Planning One of the rights bestowed under the UK government Localism Act (relating only to England) gives local groups powers to create a local development plan which, if adopted following a referendum, becomes part of the local planning framework, alongside the Local Plan.

New Economy A global meta-narrative for several initiatives or 'tribes' that aim to promote a circular economy and experiment with development and governance alternatives to engender the commons.

A. Burnett, *The Politics of Transition*, Palgrave Studies in Sub-National Governance,
https://doi.org/10.1007/978-3-031-40769-7

Relocalisation Initiatives to promote local multiplier effects and feedback loops within the local community and economy so that more activity takes place at the local level, therefore increasing resilience.

Spatial Planning Spatial planning is the policy configuration to determine planning in a given area, or multiple areas, which typically goes beyond land-use considerations to integrate factors that influence the nature and function of place.

Sustainability Transitions A normative branch of scholarship that explores how different sectors can transition towards a more sustainable and low-carbon society.

The Commons Community and natural assets managed/owned for the benefit of all and nature.

Transition Arenas A term used by sustainability transitions scholars, notably those exploring Transition Management, to describe the purposeful governance arrangements to solicit 'transitions'. It is recognised here that these may not always be purposeful, and some arenas may be subject to group capture for subversive (i.e., niche-directed) transitions.

Section 215 A Section 215 can be issued to a landowner to force remediation of a site on or adjacent to land owned by the council.

Transition (Town) Movement—Transition A global network of grassroots community projects that aim to increase self-sufficiency to reduce the potential effects of peak oil, climate destruction and economic instability.

Urban Planning The decision-making processes related to the components between different land uses (e.g., residential dwellings, employment and industry, leisure and tourism and services more broadly).

Index[1]

[1] Note: Page numbers followed by 'n' refer to notes.

A. Burnett, *The Politics of Transition*, Palgrave Studies in Sub-National Governance,
https://doi.org/10.1007/978-3-031-40769-7

O

P

SPRINGER NATURE

GPSR Compliance

The European Union's (EU) General Product Safety Regulation (GPSR) is a set of rules that requires consumer products to be safe and our obligations to ensure this.

If you have any concerns about our products, you can contact us on ProductSafety@springernature.com

In case Publisher is established outside the EU, the EU authorized representative is:

Springer Nature Customer Service Center GmbH
Europaplatz 3
69115 Heidelberg, Germany

The manufacturer's authorised representative in the EU is Springer
Nature Customer Service Centre GmbH, Europaplatz 3, 69115 Heidelberg,
Germany. If you have any concerns regarding our products, please
contact ProductSafety@springernature.com

Printed and bound by CPI Group (UK) Ltd, Croydon, CR0 4YY
01/10/2026
02225903-0012